METABOLIC ADAPTATION TO EXTRAUTERINE LIFE

DEVELOPMENTS IN PERINATAL MEDICINE

VOLUME 1

METABOLIC ADAPTATION TO EXTRAUTERINE LIFE

The antenatal role of carbohydrates and energy metabolism

Proceedings of a Workshop held in Brussels, December 19-21, 1979. Sponsored by the Commission of the European Communities, as advised by the Committee on Medical and Public Health Research.

R. DE MEYER
(editor)

1981

MARTINUS NIJHOFF PUBLISHERS

THE HAGUE / BOSTON / LONDON

for

THE COMMISSION OF THE EUROPEAN COMMUNITIES

Distributors :

for the United States and Canada
Kluwer Boston, Inc.
190, Old Derby Street
Hingham, MA 02043
USA

for all other countries
Kluwer Academic Publishers Group
Distribution Center
P.O. Box 322
3300 AH Dordrecht
The Netherlands

This volume is listed in the Library of Congress Cataloging in Publication Data

ISBN 978-94-011-7516-6   ISBN 978-94-011-7514-2 (eBook)
DOI 10.1007/978-94-011-7514-2

Publication arranged by :
Commission of the European Communities,
Directorate-General Information Market and Innovation,
Luxembourg

EUR 6858

LEGAL NOTICE

CONTENTS

VI

PREFACE

On December 19th, 20th and 21st 1979, we had the opportunity
to organize a Workshop on "Antenatal factors affecting metabolic adaptation to extrauterine life - Role of carbohydrates and energy metabolism". This meeting was made possible
thanks to grants from the Committee of Medical Research and
Public Health (CRM) of the Commission of European Communities.
We want to express our warmst gratitude for the effort that
has been made and we hope that the exchange of information
between the participants will allow better understanding of
the mechanism of adaptation to extrauterine life and that
the conclusions drawn will contribute to a better care of
the newborn.

Adaptation to extrauterine life is one of the most important steps in life. At the moment of birth, numerous
conditions are drastically changed and in order to survive
the newborn has to rely upon new sources of energy and of
substrates. This switch over has to occur smoothly and any
adaptation failure will result in the disturbance of important metabolic functions. Mortality and morbidity in neonatal period remain still very high in comparison with the
figures observed later in life; the reduction of mortality
has been much more important in child and adult than in
neonates.
Carbohydrates being the major energy supply for the fetus

and the newborn, it is obvious that this fuel has to be
considered preferentially in a meeting dealing with adaptation to extrauterine life. A lot of research work has been
done in the field of metabolic changes in the perinatal
period. However the results are sometimes conflicting.
This is partially due to the fact that experimental conditions vary from one investigator to another. On the other
hand numerous points of primordial importance remain still
unexplained.
Keeping this in mind, the aim of this Workshop was to study
the following points :
1) With regard to methodology, we want to compare the results of different investigators and to discuss their
discrepancies according to differences in their experimental conditions. Eventually this approach will draw new
lights on some factors involved in these subtile metabolic
regulations.
2) Better understanding of the factors involved in the metabolic regulation during the neonatal period must allow
some practical conclusions and rules useful for supervision
of children in the neonatal period. So, advice will be
given concerning the metabolic parameters to be monitored,
the schedule of feeding, the quantity of food, parenteral
nutrition and the effects of drugs upon the normal process
of adaptation.
3) From the discussions of the results gained by different
investigators it will be possible to evoke new hypothesis
and establish new lines for further research. It seems that
this gathering together of several people interested in
different aspects of the same fundamental question will
lead to useful exchange of information and favour new appro-
aches of the problem.

The <u>antenatal</u> situation is characterized by utilization of
high amounts of carbohydrates for energy purpose. At the
same time glucose is also transferred through the placenta
for glycogen storage and for transformation to lipids.
There is no - or nearly no - glyconeogenesis from amino
acids which are used exclusively for protein synthesis.
The moment at which, during maturation, the capacity of
elaborating glycogen and lipid stores appear is variable
from one species to the other, but all species do elaborate
stores. This phenomenon is of primary importance to cover
the energy needs during the early postnatal period. This
particular metabolic situation of continuous anabolism is
made possible by the fact that the fetus is "grafted" upon
the maternal circulation and that all the metabolites are
offered to him among which he can select material according
to his needs.
What characterizes the <u>postnatal</u> period is in a first stage
the total <u>absence of any supply</u> and in a second stage <u>oral</u>
<u>nutrition</u>. During the early postnatal period when the um-
bilical cord is severed but before the first feeding the
machinery keeps still working for some time in an anabolic
way. This means that glycogen is still synthesized from
glucose but the lack of exogenous supply is responsible for
the exhaustion of blood stream substrates such as glucose
and amino acids. At that time, the glycogen stores in the
liver starts delivering glucose. This restores the normal
glucose level, and keeps normal metabolism on drive in the
tissues specially the brain which is more or less dependent
on the glucose supply. Shortly after, glyconeogenesis
starts. The amino acids still in the blood or issued from
the normal catabolism of proteins undergo transformations
to glucose. Once the first feeding has taken place, the
newborn proceeds to the next stage, characterized by :
1) a discontinuous supply of nutrients; 2) the lack of
choice among the metabolites offered. They have to be
accepted in bulk as they enter the organism through the
intestinal tract.
This situation makes the interconversion of the substrates

absolutely necessary. Indeed if no interconversion is possible, the needs will not correspond qualitatively to the
substrates offered and there will be an excess of some substrates not used. This is a major difference in comparison
to the antenatal situation when the fetus having selected
the substrates he needs, the excess was returned to the
maternal circulation. In this new situation, the newborn
cannot keep constant the aminoacidogram in the blood without a mechanism favouring interconversion. So the two most
important capabilities during the neonatal period are the
mobilization of the stores from carbohydrates and from
lipids and the interconversion of substrates principally
glyconeogenesis. This adaptation machinery is induced by
different mechanisms.
1) Enzymes for all the metabolic steps have to be present.
Some of these are present already during intrauterine life.
Others are induced after birth. They have to be in an
active form. Sometimes enzymes are present very early before birth, but the switch over from an inactive to an active form happens only at birth.
2) Hormones play likely a role in the induction and activation for the enzymes. Corticoids are necessary for the
synthesis of glycogen : alterations in the insulin to glucagon ratio is important in the induction of gluconeogenesis.
Catacholamines are factors in glycogenolysis. Likely other
hormonal factors are also active, for instance, placental
factors acting on insulin release by the fetus, thyroid
hormones acting on maturation of the sensitivity to hormones. An important point for the potential activity of hormones is the presence of receptors. A major problem is to
know which factors are able to elicit these receptors. In
this regard, thyroid hormones play perhaps an important
role.
Going one step further, one may wonder why the hormonal situation is changed postnatally. Among the possible factors
which are involved we keep in mind hypoglycemia, oxygen
supply to the tissues and specially to the liver, environmental factors : temperature, light, mechanical pressure,

the disappearance of placental function with accumulation
of metabolites commonly withdrawn through the placenta.

Better knowledge of these mechanisms will lead to better
understanding of what happens in abnormal conditions : pre-
maturity, dysmaturity, maternal diabetes and hypothyroidism,
administration of drugs to the mother, feeding troubles of
the newborn. It must thus be possible to foresee different
guide lines for practical recommendations in the management
of the newborn. However, the presentation and the discussion of the last session of the Workshop have shown that it
is too early to draw definite conclusions and practical
recommendations.
Some points nevertheless do merge from this session.
1) The advantages of early feeding, specially when the
stores are mobilized with difficulty or are inexistant.
2) The necessity to give lipids in early feeding.
But many points still remain unclear, such :
a) What is the best feeding for the premature?
b) How to manage exactly parenteral nutrition? It is likely that in the next few years there will be a better approach to these problems. It is therefore likely that it will
be useful in two or three years to bring together the people
interested in this fiels and to analyse the problems from
a more practical point of view. The topic of the feeding
of the newborn could be excellent for another Workshop.
As far as the lines for further research are concerned, I
just want to mention :
a) The better analysis of the role of the placenta in the
prenatal preparation to extrauterine life.
b) A better understanding of what happens in the very early
neonatal period (first two hours). One of the most puzzling
questions in this regard : where is glucose coming from,
during this time when there is no gluconeogenesis, nor
glycogenolysis? What are the factors that can induce an
earlier gluconeogenesis or inhibit its normal onset (effect
of drugs)?

c) Another very interesting point is to know what will be
the effects of a bad adaptation or of drugs administration
during the neonatal period in later life.

It has been well documented that in the neonatal period,
several control mechanisms are established (development of
the hypothalamus towards a male or female type, determination of normal blood glucose level, determination of blood
pressure level). At that moment, the organism is very
sensitive to any disturbance and permanent anomalies can be
induced. So for instance in the rat, treatment with gona-
dal steroids influences the function of the hypophysis at
puberty. Giving angiotensin modifies the arterial pressure
in later life. Thyroid hormone given during the neonatal
period will modify the TSH secretion, in adults.

d) Another yet unresolved problem is to determine what are
the factors which will modify in the neonatal period the
sensitivity to certain hormones, for instance to glucagon,
insulin. Here one must evoke the problem of factors inducing new receptors.

Finally, from a methodologic point of view, different people being involved in problems closely related to each
other had the opportunity to compare their results in different species and to elucidate some of the discrepancies
by looking more closely of the differences within experimental conditions. So, for instance, it became clear that
hypoglycemia in the neonatal period is highly dependent on
the environmental temperature; therefore it is important to
take in account the incubation temperature if one compares
the different experiments. On the other hand, the influence
of intrauterine growth retardation on glycogen is entirely
different according to the species considered : rat, rabbit,
or guinea pig. It could be very interesting to look closer
of these discrepancies and try to find out why the different species behave differently.

We do hope that the information brought together in this Workshop will be of benefit to the too numerous newborns who are still unadequately adapted when they go through the most hasardous event of their life, that is birth.

LIST OF PARTICIPANTS

BELGIUM
DE MEYER R.
Laboratoire de Tératologie et de Génétique Médicale
Université Catholique de Louvain
U.C.L. 5350 - Tour Pasteur
avenue E. Mounier, 53
1200 BRUXELLES

DEROM R.
Verloskundige Kliniek
Academisch Ziekenhuis
Rijksuniversiteit Gent
De Pintelaan 135
9000 GENT

DEVOS P.
Laboratoire de Biochimie
Université Catholique de Louvain
U.C.L. 7539 - I.C.P.
avenue E. Mounier, 75
1200 BRUXELLES

EGGERMONT E.
Neonatologie
Kinderkliniek "Gasthuisberg"
Katholieke Universiteit Leuven
Herestraat 49
3000 LEUVEN

HERS H.G.
Laboratoire de Biochimie
Université Catholique de Louvain
U.C.L. 7539 - I.C.P.
avenue E. Mounier, 75
1200 BRUXELLES

SENTERRE J.
Service de Pédiatrie
Université de Liège
Hôpital de Bavière
4020 LIEGE

SODOYEZ F.
Service de Pédiatrie
Université de Liège
Hôpital de Bavière
4020 LIEGE

VAN ASSCHE A.
Dienst Verloskunde
St. Rafaëlkliniek
Katholieke Universiteit Leuven
Kapucijnenvoer 33
3000 LEUVEN

VERELLEN G.
Service de Néonatologie - Département de Pédiatrie
Cliniques Universitaires Saint-Luc
Université Catholique de Louvain
avenue Hippocrate 10
1200 BRUXELLES

DENMARK

GREGERSEN N.
Research Laboratory for Metabolic Disorders
University Department of Clinical Chemistry
Aarhus Kommunehospital
8000 AARHUS C

RASMUSSEN K.
Research Laboratory for Metabolic Disorders
University Department of Clinical Chemistry
Aarhus Kommunehospital
8000 AARHUS C

FRANCE

BERTRAND J.
Unité de Recherches endocriniennes et métaboliques chez
l'enfant
INSERM - U 34
Hôpital Debrousse
69322 LYON CEDEX 1

FERRE P.
Laboratoire de Physiologie du Développement
Collège de France
Place Marcelin Berthelot 11
75231 PARIS CEDEX 5

GIRARD J.
Laboratoire de Physiologie du Développement
Collège de France
Place Marcelin Berthelot 11
75231 PARIS CEDEX 5

MINKOWSKI A.
Centre de Recherches de Biologie du Développement Foetal
et Néonatal
INSERM - U 29
Hôpital de Port Royal
Université René Descartes
Boulevard de Port Royal 123
75674 PARIS CEDEX 14

ROUX J.M.
Centre de Recherches de Biologie du Développement Foetal
et Néonatal
INSERM - U 29
Hôpital de Port Royal
Université René Descartes
Boulevard de Port Royal 123
75674 PARIS CEDEX 14

SALLE B.
Service de Néonatologie et de Réanimation Néonatale
Pavillon J
Hôpital Edouard Herriot
Place d'Arsonval
69374 LYON CEDEX 2

GERMANY

HELGE H.
Kinderklinik und Poliklinik
Universitätsklinikum Charlottenburg
FUB Heubnerweg 6
1000 BERLIN 19

TELLER W.
Departement für Kinderheilkunde der Universität Ulm
Prittwitzstrasse 43
Postfach 3880
7900 ULM/DONAU

IRELAND

MOORE R.E.
Department of Physiology
Trinity College
University of Dublin
DUBLIN II

ISRAEL

MOSES S.W.
Pediatric Research Laboratory
Soroka Medical Center - School of Health Services
Ben Gourion University of the Negev
BEER SHEBA (P.O.B. 151)

THE NETHERLANDS

KOLLEE L.A.A.
Department of Pediatrics
Sint Radboudziekenhuis
Katholieke Universiteit Nijmegen
Geert Grooteplein Zuid 20 - Postbus 9101
6500 HB NIJMEGEN

KOPPE J.G.
Department of Physiology and Pathology of the Newborn
Wilhelmina Gasthuis
1e Helmersstraat 104
1054 EG AMSTERDAM

SAUER P.J.J.
Department of Pediatrics
Academic Hospital Rotterdam
Sophia Children's Hospital and Neonatal Unit
Gordelweg 160
3038 GE ROTTERDAM

SWITZERLAND

BOSSI E.
Inselspital Bern
Medizinische Universitäts-Kinderklinik und Poliklinik
3010 BERN

COLOMBO J.P.
Inselspital Bern
Chemisches Zentrallabor der Universitätskliniken
3010 BERN

SIEGRIST H.
Inselspital Bern
Medizinische Universitäts-Kinderklinik und Poliklinik
3010 BERN

UNITED KINGDOM

BRITTON H.G.
Department of Physiology
St. Mary's Hospital Medical School
University of London
PADDINGTON, LONDON W2

DUFF D.A.
Department of Biochemistry
University of Surrey
GUILFORD, SURREY GU2 5XH

HOOTON-SHELLEY H.J.
Nuffield Institute for Medical Research
University of Oxford
Headley way
HEADINGTON, OXFORD OX3 9DS

HULL D.
Medical School - Department of Child Health
Queen's Medical Centre
University of Nottingham
Clifton Boulevard
NOTTINGHAM NG7 2UH

JONES C.T.
Nuffield Institute for Medical Research
University of Oxford
Headley way
HEADINGTON, OXFORD OX3 9DS

MILNER R.D.G.
Department of Paediatrics
Children's Hospital
University of Sheffield
SHEFFIELD S10 2TH

SNELL K.
Department of Biochemistry
University of Surrey
GUILFORD, SURREY GU2 5XH

SESSION I
CARBOHYDRATE UTILIZATION
IN THE NEWBORN

THE METABOLISM OF GLYCOGEN IN THE LIVER OF FOETAL AND NEWBORN RATS

H.G. Hers

Glycogen is known to accumulate slowly in the liver during the last days of gestation and to be degraded soon after birth. In this presentation, I will briefly describe some peculiar properties of the synthesis of foetal glycogen and of its post-natal degradation.

THE SYNTHESIS OF FOETAL LIVER GLYCOGEN

The concentration of glycogen in the liver at birth reaches 8 to 10 %, a value which is at least twice higher than the usual concentration of glycogen in the adult liver. This high concentration is however not the result of a rapid synthesis of glycogen during the last days of gestation. Indeed, this synthesis occurs over a period of about 48 hours, and the mean rate of synthesis during this period remains several fold below the maximal rate of 1 % per hour reached in the liver of adult rats upon refeeding after fasting. This relatively slow rate of synthesis is accounted for by a corresponding activity of glycogen synthase in its active form (synthase a). If glycogen accumulates under these conditions, it is because contrary to what is observed in the adult, there is no nychthemeral variation of the glycogen content in the foetal liver. Accordingly, we have been able to show that the radioactive glycogen formed in the first period of synthesis during the foetal life is still present in the liver at birth, indicating a near absence of glycogen degradation (1).

Another remarkable feature of this synthesis of
glycogen was that radioactive glucosyl units incorporated
into glycogen at day 19.5 of gestation (i.e. at the beginning of the synthesis of glycogen) were almost equally
distributed on the inner and outer chains of the poly-
saccharide at birth; these radioactive units had therefore
not been covered by the unlabelled glycogen which was
formed in large amount during the 2 last days of gestation
(1). This indicates that some glycogen molecules are build
up completely before other ones are synthetized. This
characteristic of glycogen synthesis has now been found
also in the adult liver (2) and appears to be related to
a general property of glycogen synthase (3).

POST-NATAL GLYCOGEN DEGRADATION

There are several indications in the literature that the
post-natal glycogenolysis in the liver may not be phosphorolytic but could occur by hydrolysis inside of autophagic
vacuoles. Indeed, Snell and Walker (4) have shown that the
glycogen present in the liver of new-born rats is relatively resistant to glucagon administration. On the other hand,
Jézéquel et al. (5) have observed under the electron micro-
scope the presence of numerous glycogen-filled vacuoles in
the neonatal mouse liver. This observation has been confirmed by Kotoulas and Phillips (6) who have also reported
that the administration of puromycin to new-born rats prevented almost completely the formation of glycogen-filled
autophagic vacuoles in the liver, although it intensified
glycogenolysis. Puromycin is indeed known to activate
phosphorylase (7) and to be glycogenolytic in the adult
liver (8).

The existence of a hydrolytic and lysosomal pathway
of glycogen degradation is known by the work of Hers (9)
on type II glycogenosis. Indeed, in this disease, there
is a complete deficiency of a lysosomal α-glucosidase and
an intravacuolar accumulation of glycogen (10). The role

of this hydrolytic mechanism is believed not to provide
glucose to the cell but rather to destroy the glycogen
which has been degraded in the vacuolar system during the
process of autophagy (9). Glycogen-filled lysosomes are
very rarely seen in a normal liver and their presence in
the livers of new-born rats is highly suggestive of a participation of lysosomes in glycogen degradation in this
very specific period of life.

It is only very recently that a biochemical approach
to this problem has been made possible. Indeed, we have
been able to show that glycogen degradation in the adult
liver follows a molecular order in this respect that the
glucosyl units that have been incorporated last during
synthesis are removed first and vice versa (2). This
ordered degradation cannot be explained by a preferential
removal of the outer chains but is the reflect of an ordered synthesis of glycogen molecules and of their degradation in the reverse order. This ordered mechanism is a
property of the glycogen molecule itself and of the enzymes
involved in its synthesis and degradation. Indeed, the
ordered degradation can be observed in a cell-free system
upon incubation in the presence of inorganic phosphate of
freshly isolated particulate glycogen (2) to which phosphorylase and other glycogenolytic enzymes remain closely
associated during the isolation procedure.

One can expect that this ordered degradation would not
be observed in an autophagic and hydrolytic mechanism.
Indeed, autophagy is believed to occur at random in the
cell and there is no indication that the hydrolysis of
glycogen by α-glucosidase could occur orderly. Accordingly,
we show in fig. 1a that glycogen which has been labelled
during the initial period of synthesis in the foetal liver
(19.5 day of gestation) is degraded at the same rate as
the total mass of glycogen, i.e. at random. In contrast,
fig. 1b shows that after puromycin administration to the
new-born rats, this "early labelled" glycogen started to
be degraded after a two hours delay. This pattern of
degradation is characteristic of phosphorolytic degradation,

6

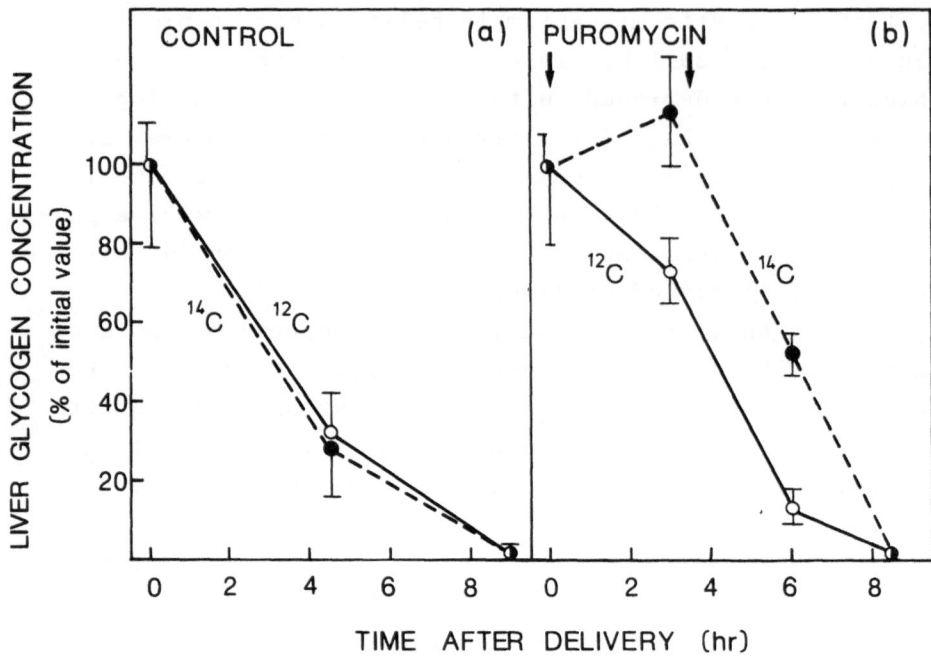

<u>Figure 1</u> : <u>Degradation of glycogen in the livers of</u>
<u>new-born rats treated or not with puromycin</u>

The glycogen was made radioactive by injection of [U-^{14}C]
glucose to the mother at day 19.5 of gestation. The
foetuses were delivered by cesarian section at day 21.5.
Puromycin (0.4 mg) was injected intraperitoneally at time
zero and 3 hrs later (arrows). The animals were maintained at 37°C in moistened air and killed at the time
indicated. Their livers were weighed and dropped in 1 ml
of hot 5 N KOH for the isolation of glycogen. The concentration of glycogen at time zero was equal to 8.1 %
in fig. 1a and 6.2 % in fig. 1b . The specific radioactivity approximated 1,000 cpm mg^{-1} in both cases.

as was observed in the adult animal (2). This observation
allows to conclude that the degradation of glycogen which
occurs in the liver of new-born rats during the first
hours after birth, proceeds by a mechanism which is different from the normal phosphorolytic degradation.

A similar parallelism between a random degradation of
glycogen in the presence of autophagic vacuoles has also
been observed in the liver of adult rats treated with

phlorizin (3). Such a parallelism is suggestive of an auto-phagic and lysosomal degradation of glycogen by acid α-glucosidase. It must be recognized, however, that the amount of α-glucosidase currently measured in the liver (0.3 unit/g) appears not to be quite sufficient to account for the high rate of glycogenolysis reported in fig. 1a (close to 1 μmole of glucose/min^{-1}.g^{-1}). One would have therefore to assume that the activity of the enzyme *in vivo* may be, by an unknown mechanism, greater than that measured *in vitro*.

REFERENCES

1. Devos, P. and Hers, H.G., Glycogen metabolism in the liver of the foetal rat. Biochem. J. 140:331-340, 1974.
2. Devos, P. and Hers, H.G., A molecular order in the synthesis and degradation of glycogen in the liver. Eur. J. Biol. Chem. 99:161-167, 1979.
3. Devos, P. and Hers, H.G., in preparation.
4. Snell, K. & Walker, D.G., Glucose metabolism in the new-born rat. Hormonal effects *in vivo*. Biochem. J. 134:899-906, 1973.
5. Jézéquel, A.-M., Arakawa, K. & Steiner, J., The fine structure of the normal, neonatal mouse liver. Lab. Invest. 14:1894-1930, 1965.
6. Kotoulas, O.B. and Phillips, M.J., Fine structural aspects of the mobilization of hepatic glycogen. I. Acceleration of glycogen breakdown. Amer. J. Pathol. 63:1-17, 1971.
7. Hofert, J.F. & Boutwell, R.K., Effect of puromycin on hepatic glycogen phosphorylase activity. Proc. Soc. Exptl. Biol. Med. 121:532-536, 1966.
8. Hofert, J.F. & Boutwell, R.K., Puromycin : a potent metabolic effect independent of protein synthesis in mouse liver : effects of puromycin analogs. Arch. Biochem. 103:338-344, 1963.
9. Hers, H.G. α-Glucosidase deficiency in generalized glycogen storage disease (Pompe's disease). Biochem. J. 86:11-16, 1963.
10. Baudhuin, P., Hers, H.G. & Loeb, H., An electron microscopic and biochemical study of Type II glycogenosis. Lab. Invest., 13, 1139-1152, 1964.

HORMONAL CONTROL OF FETAL LIVER GLYCOGEN METABOLISM

S.W. Moses, M.D., M. Pines, M.Sc., and N. Bashan, Ph.D.

INTRODUCTION

Claude Bernard, 120 years ago, was the first to observe glycogen accumulation in the mammalian liver during fetal development. Similar observations have since been extended to all mammalian species examined. However, the timing and the extent of accumulation differ among species: e.g. in the rat liver glycogen begins to accumulate on day 18 of gestation (1), whereas in the human fetus glycogen deposition starts during the first trimester and continues to accumulate throughout gestation (2).

Glycogen accumulation requires the presence of glycogen synthetic enzymes, their respective co-factors and the necessary hormonal profile for their function. In view of the differences in timing of glycogen accumulation between species, it is to be expected that both enzyme activities and hormone levels also show species related specificities. Most of our information stems from rat liver homogenates and more recently liver explant and hepatocyte culture. Only little information is available about the human fetal liver glycogen development and its hormonal controls.

CORTISONE AND INSULIN EFFECT ON GLYCOGEN ACCUMULATION

In rat liver homogenates glycogen begins to accumulate on day 18 to reach a peak of 70 mg/g liver on day 21 before the abrupt perinatal drop which occurs between the 21st and 22nd days of gestation (3). This glycogen synthetic deposition has an obligatory requirement for cortisone. Glycogen synthesis was compromised if both maternal and fetal steroid production was prevented by adrenalectomising the mother and hypophysectomising by decapitation or adrenalectomising the fetus on day 18 of gestation. The administration of cortisone restored fetal rat liver glycogen content (Fig. 1) (4). It has since been established that glycogen synthetase, which increased sharply on day 18 in the normal rat, was suppressed in animals deprived

of corticosteroids and was restored to normal by the administration of cortisone (5).
Greengard and Dewey demonstrated that the injection of hydrocortisone into rat
fetuses caused a premature and enhanced glycogen deposition in their livers (6).
Pines (7) showed that injection of hydrocortisone 24 hours before the injection of
radioactive glucose C^{14}, sacrificing the fetuses one hour later, incorporation rates
started earlier and reached a peak which was 63% higher in hydrocortisone administered rats as compared to controls.

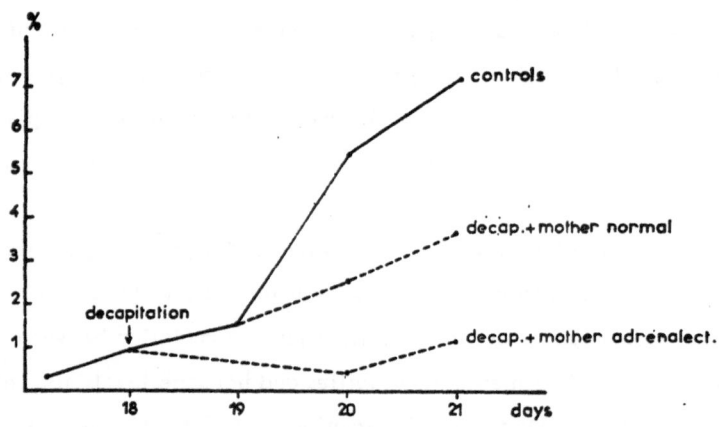

Fig. 1: Effect of endogenous steroids on fetal rat liver glycogen synthesis.

Rat fetuses: glycogen content of the liver (percentage of fresh tissue)
in control fetuses, and in fetuses decapitated on day 18 either in normal mother animals or in adrenalectomized mother animals. Modified
after Jacquot (1959).

On day 21 both injected animals and controls showed a similar trend of decrease, which however, in the treated animals, remained three times higher than
in controls (Fig. 2).

Hydrocortisone caused an elevation of glycogen synthetase a + b activity
which started on day 17 and was most marked on days 18 and 19 (Fig. 3). In contrast hydrocortisone caused an impressive enhancement of glycogen synthetase a
activity from day 18 onwards, which continued during the whole last period of
gestation (Fig. 4). Vanstapel showed that corticosteroid administration induced
the development of glycogen synthetase phosphatase (9). On the other hand, it

is quite evident that on days 17 to 19 the injection of hydrocortisone had no effect on glycogen phosphorylase a activity, whereas a slight decrease was noted on days 20 and 21 (Fig. 5) (7).

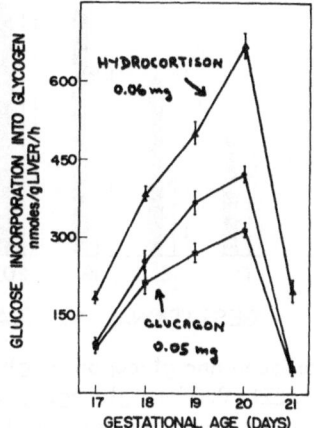

Fig. 2: Effect of hydrocortisone and glucagon injection in fetal rats on glucose-C[14] incorporation rates into glycogen in their livers.
Hormones were injected 24 hours before glucose-C[14]. Animals were sacrificed 1 hour after. ▲ 0.06 mg hydrocortisone (I.P.); ■ 0.05 mg glucagon (I.P.); ● Controls.

Eisen compared the hydrocortisone effect in vivo with the effect in liver explant culture and found a similar pattern of induction of total glycogen synthetase (Fig. 6). This induction was suppressed in the presence of Actinomycin D (8). A similar requirement for cortisol can be elicited in fetal hepatocyte culture. Cells, cultivated on day 14, when provided with cortisol in their incubations medium started to accumulate glycogen on day 15 (10). In contrast when subsequently cultivated in cortisol free medium, they started to lose their glycogen. It is thus evident that liver cells are competent to permit induction of glycogen synthetase at least three days before the onset of the physiological induction of the enzyme and that the presence of cortisone is essential for the deposition of glycogen. In the human fetus cortisol is secreted by the fetal adrenal from an early age, plasma concentrations rise toward term to fall thereafter (11). It is expected that cortisol in the human fetus has a similar function to the one observed in the rat fetus.

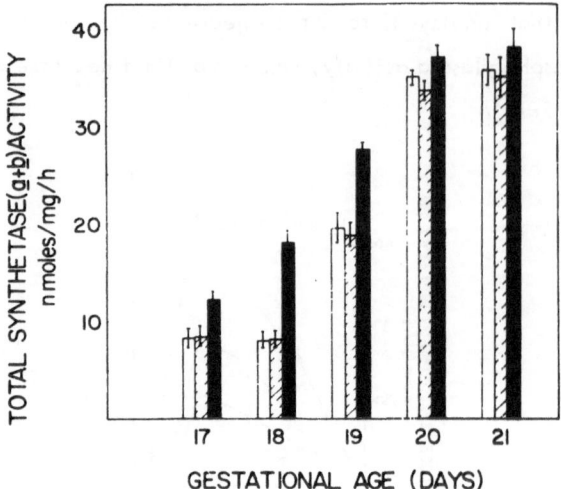

Fig. 3: Effect of hydrocortisone and glucagon on glycogen synthetase a + b in fetal rat livers. ▨ 0.05 mg glucagon, ▬ 0.06 mg hydrocortisone, ▢ control hormones were injected I.P. into rat fetuses: hyrocortisone 24 hours and glucagon 1 hour before sacrificing the animals.

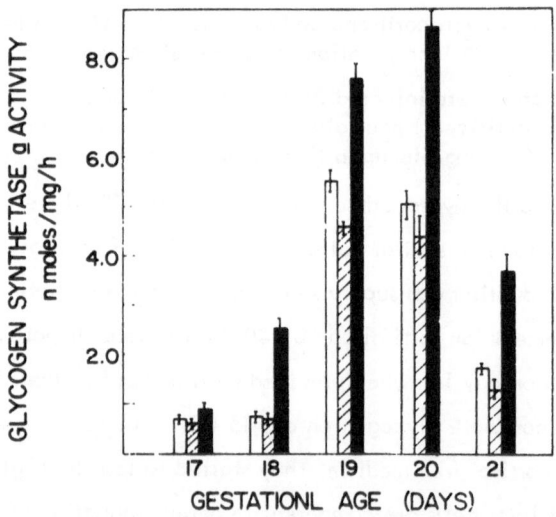

Fig. 4: Effect of hydrocortisone and glucagon on glycogen synthetase a activity in fetal rat livers. ▨ 0.05 mg glucagon, ▬ 0.06 mg hydrocortisone, ▢ control hormones were injected I.P. into rat fetuses. Hydrocortisone 24 hours and glucagon 1 hour before sacrificing the animals.

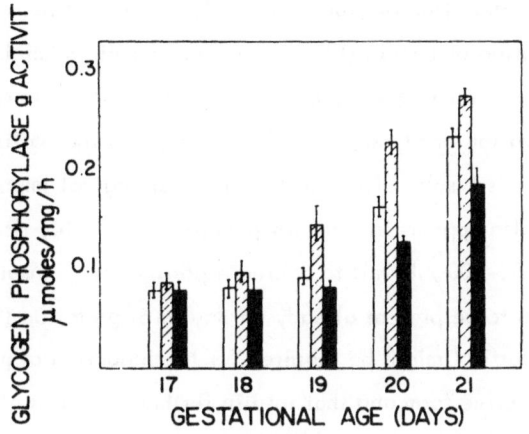

Fig. 5: Effect of hydrocortison and glucagon on the activity of glycogen phosphorylase a in fetal rat livers. ▨ 0.05 mg glucagon, ■ 0.06 mg hydrocortison, ☐ control, hormones were injected I.P. into rat fetuses: hydrocortisone–24 hours and glucagon 1 hour before sacrificing the animals.

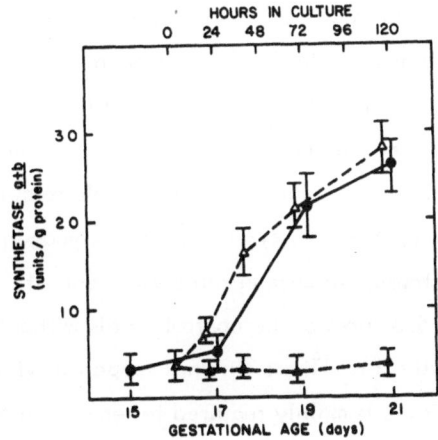

Fig. 6: Development of fetal rat liver glycogen synthetase a + b in utero and in vitro. ● in utero; △ in vitro + 10 µM hydrocortisone; ▲ in vitro control;
From Eisen et al, 1973.

In the fetal rat hepatocytes, induction of glycogen synthetase by cortisol causing increased incorporation of glucose into glycogen, can be augmented by subsequent administration of insulin (12). However, insulin administration without cortisone does not result in glycogen accumulation. Thus the onset of an insulin dependent activation of glycogen synthetase requires the earlier permissive action of cortisol. This insulin effect may have physiological importance as immuno-reactive insulin appears in fetal rat pancreas on day 14, rises at term to high concentrations (50–200 μU/ml) to decrease postpartum. Insulin binding sites have been shown to be present already on day 16 of gestation (8). It has been suggested that cortisol might be required for the induction of glycogen synthetase mostly in its active form and that insulin further increases glycogen synthetase activity in its active form. The insulin effect is apparently partly related to protein synthesis as cycloheximide inhibits this effect (13).

Plas recently described a time dependent insulin effect in fetal rat hepatocytes. He found that 4 hours after exposure of hepatocytes to insulin, a resistance developed in these cells so that subsequent exposures resulted in a decreased effect both in terms of glucose incorporation into glycogen and glycogen content. This resistance was dose dependent. The higher the initial insulin concentration the stronger the resistance. No interference with receptor binding has been found. The presence of an intracellular inhibitor has not yet been ruled out (14). A final explanation has not yet been given and the importance of this feature in the actual regulation of fetal liver glycogen metabolism has not yet been established.

Eisen found that in liver explants taken from 21-day-old term fetuses – hydrocortisone did not affect total synthetase, synthetase a or glycogen levels, whereas insulin without cortisone increased glycogen synthetase a activity 3.5 times, causing glycogen to accumulate 5.5 times of the control levels without affecting glycogen synthetase a + b activity (8). These findings support in vivo studies presented by M. Pines that cortisone is mainly required for enzyme induction. Once enzyme activity has received its full expression cortisone has no effect. In contrast insulin at this period, is mainly involved in enzyme interconversion leading to a rise in glycogen synthetase a activity.

Insulin is detectable in the human fetal pancreas in the eighth week of gest-

ation.   Schwartz et al showed that insulin was able to increase glycogen concentration in a thirteen-week-old human fetal liver explant, however, pharmacological doses were required (1.0 U/ml) and the effect was very slow (Fig. 7).   The
insulin effect was due to a small increase in synthetase a activity.   Phosphorylase
activity was not affected by this treatment (2).

Addition	Glycogen μg Glucose/mg Protein (n = 5)
None	$35,14 \pm 2,9$
Glucagon 7,5 µg/ml	$14,21 \pm 0,8$
Insulin 1 U/ml	$45,05 \pm 6,1$
Glucagon + Insulin	$14,93 \pm 0,3$

Fig. 7: Effect of glucagon and insulin on the glycogen concentration in human
        fetal liver explant (gestational age 16 weeks).   Glycogen concen-
        tration was measured 6 hours after the addition of the hormones.
        Adapted from A.L. Schwartz, 1975.

In summary, in the fetal rat liver cortisone is required for the induction of glycogen synthetic enzymes.   It is capable both of causing premature induction and
enhancing glycogen synthesis.   Glycogen deposition is further augmented by insulin.   Insulin affects mainly the activation of glycogen synthetase, however, a
protein synthetic process cannot be ruled out.   In the human glycogen accumulation starts early in gestation.   No data on cortisone effect are available.   The
insulin effect is in a similar direction as in the rat, however, the sensitivity to
insulin is very low.

GLUCAGON EFFECT ON GLYCOGEN METABOLISM

Glycogen phosphorylase activity is already present on day 16,5 of gestation in the rat (1,7) to rise gradually to reach adult values at term. A more or less parallel pattern was observed in respect to phosphorylase a which comprised about 80% of the total enzyme (15). In vivo studies, measuring the turnover of glycogen suggested that fetal phosphorylase is mainly inactive (1).

Pines showed that in contrast to the effect of hydrocortisone, which caused a precocious and higher rate of glucose incorporation into glycogen, glucagon had an opposite effect (Fig. 2) decreasing incorporation rates by 35% on day 19. As changes in incorporation rates do not distinguish readily between synthesis and breakdown, the glucagon effect shown here may represent inactivation of glycogen synthetase or activation of glycogen phosphorylase. It is evident from Fig.5 that in vivo a direct glucagon effect on phosphorylase a activity was not present before day 19 which decreased towards day 21. In contrast glucagon had little effect on total glycogen synthetase (Fig. 3) or glycogen synthetase a (Fig.4) (7). The effect of glucagon on increasing phosphorylase a and of decreasing glycogen synthetase a activity is of a small magnitude as compared to the adult animal. Injections of glucagon in vivo resulted in the adult in a 78% decrease in activity of glycogen synthetase (16) as compared to 17% in the fetus (7). The effect of the hormone on adult phosphorylase, measured in isolated livers, rises twofold and in liver slices by 80% (17). Glucagon caused an increase in glycogen phosphorylase a of 31% in the 20-day-old fetus (7). The small sensitivity of these enzymes to glucagon during the last days of gestation are in accord with the studies of Schaub and Becker who showed that activation of phosphorylase by the administration of glucagon started antenatally on day 19 of gestation, whereas glucagon stimulation caused inactivation of glycogen synthetase only postnatally (18).

As this action of glucagon on glycogen metabolism is by activation of adenyl cyclase, which increases cyclic AMP levels, Pines studied the question whether the moderate effect of glucagon was related to a limiting effect of glucagon on adenyl cyclase. In vitro studies in liver homogenates showed that cyclic AMP rises as a function of gestational age and that glucagon elicits a threefold rise on day 17 and a fourfold rise on day 20 of gestation which is impressive, yet in no proportion to the tenfold rise observed in the adult rat liver (Fig. 8) (19). These data are in accord with Butcher and Potter (20).

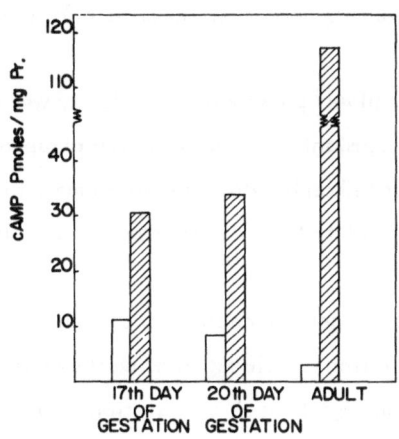

Fig. 8: Effect of glucagon on cyclic AMP levels in rat fetal liver slices. Liver slices were incubated for 20 minutes at 37° C in the presence and absence of 50 µg glucagon. ▨ + glucagon ▢ - glucagon

Prenatal protein kinase activity is higher than postnatal. In contrast phosphorylase kinase shows a different developmental pattern. Antenatal activity which is very low, increases as a function of gestational age and continues to rise after term (21). It may well be that during the last days of gestation this enzyme becomes rate limiting in terms of the glucagon effect on phosphorylase activation. Another possibility was considered by de Vos and Hers, who, in view of the lack of activation of phosphorylase b by glucagon, and a lack of inactivation of phosphorylase a by glucose suggested that fetal phosphorylase may at this age be different from the adult form (1). In contrast to these experiments, a somewhat different response has been found by Plas while studying rat hepatocyte cultures. Hepatocyte cultured on day 15 of the gestational age showed a marked glycogeno-

lytic response to 10 /uM glucagon.   He subsequently observed a pattern of resistance of glycogenolysis to additional glucagon administration four hours after
the first exposure of the hepatocytes to glucagon.   The lack of response rests
probably at the level of cyclic AMP modulation.   It is of interest to note, that
even in the presence of the refractoriness to glucagon, norepinephrine was able to
induce a glycogenolytic response.   It is as yet unclear, whether this mechanism
of resistance to glycogenolysis may be one additional factor contributing to the
maintenance of high hepatic glycogen stores before birth (12).   It is, however, of
interest to note that a refractory condition exists both for the insulin and the glucagon effect.

In human fetal liver explants phosphorylase activity was detected during the
first trimester and a small constant increase in active phosphorylase as a function
of age was found, while total phosphorylase increased more rapidly.   Phosphorylase phosphatase was also detected at an early stage of development (15-16 weeks)
(2).

Hormonal effects on glycogen concentration of human fetal liver can be
shown already in early gestation.   Glucagon as well as insulin are detectable in
human fetal pancreas by the eighth week of gestation (22).   Both hormones do not
cross the placenta.   Glucagon effect on liver glycogen was already present during the first trimester.   Glucagon decreased liver glycogen within six hours to
40% and insulin, which in high concentration caused glycogen to increase, did
not counteract the glucagon effect (Fig. 7).   The administration of glucagon and
theophylline caused one-tenth the elevation of cyclic AMP present in the adult.
On the other hand, the administration of dibutyryl cyclic AMP caused rapid degradation of glycogen suggesting that cyclic AMP (adenyl cyclase ?) is the rate
limiting step in response to glucagon.   Dibutyryl cyclic AMP had no effect on
phosphorylase activity in human fetal explants.   The decrease of glycogen concentration in the presence of unaltered phosphorylase activity has been attributed
by the author to be caused by inactivation of glycogen synthetase (2).   A similar
mechanism has been suggested to be present in the fetal rat (15).

Summarising the glucagon effect in fetal rat it is striking that on comparing
response to glucagon in hepatocytes with in vivo experiments, it is apparent that
in hepatocytes a glucagon dependent pathway is present early in development, be-

fore any net storage of glycogen occurs. In contrast to insulin, the glucagon effect does not require the prior permissive action of cortisone. In contrast in vivo
the classical Sutherland type activation of glycogen phosphorylase retains certain
rate limiting steps till very late in gestation. It is unclear whether adenylate
cyclase (23), phosphorylase kinase or phosphorylase a are the critical enzymes,
which, through late induction, limit the activity of this pathway. Girard demonstrates the presence of an activating mechanism of phosphorylase close to term
which is apparently responsible to a large extent for the rapid and massive glycogenolysis which occurs during the perinatal period (24).

The developmental pattern of other non-cyclic AMP dependent pathways of
phosphorylase activation, known to exist in the adult animal have not been studied extensively. They may be involved in the rapid conversion of phosphorylase
during traumatic killing of rat fetuses (1). In the human fetus the glycogen synthetic and degradative pathway seem to operate early in gestation and to respond
to glucagon or cAMP. The same problem of inducing glycogenolysis without
activating phosphorylase is also evident in the human species (2). Again as found
in insulin responses, basic similarities between glycogen metabolism in the rat and
in the human are evident, yet the time table and the magnitude are different.

PERINATAL CHANGES

The reverse trend in glycogen metabolism, which occurs during the perinatal period, in which a slowly progressive anabolic process is suddenly converted into an
abrupt state of glycogen breakdown, requires major changes in enzymatic controls.
These changes are apparently triggered by a combination of interreacting factors
such as neuroendocrine stimuli,lack of substrate availability and subsequent hormonal changes.   Catecholamines such as norepinephrine excretion is expected to
occur as a result of the stress the newborn undergoes during delivery.   Anoxia can
potentiate this effect (24).   Adrenalin or cortisol from the adrenal does not seem
to be a major factor, since adrenalectomy in newborn rats inhibited glycogen
breakdown only by 15 - 20% (25).   Endogenous steroid excretion does not seem
to have an important role in controlling these changes, as Dimarco et al (26) did
not find any change in free plasma or liver cortisol in postpartum rats delivered on
term.

The sudden discontinuation of the maternal glucose supply at the time of delivery leads to a marked hypoglycemia in the newborn rat.   It is well documented
(25) that blood glucose concentrations decrease immediately after birth.   Bashan
et al showed a decrease from 48 mg% to 5 mg% to recover to its former level within one hour (Fig. 9)(28) At the same time a reversal of the insulin glucagon ratio
is observed (Fig. 10)(29).   This reversal of the insulin glucagon ratio per se is not
the only factor for glycogenolysis, as it has been shown that if insulin is administered no change in phosphorylase is noted,yet, hypoglycemia and partial glycogenolysis still ensue (25).   The stimulus for pancreatic insulin and glucagon secretion at this stage has not been completely elucidated.   Newborn α and β
cells are remarkably unresponsive to changes in glycemia (30).   It is more likely
that the β cells respond to changes in plasma amino acid present in the fetus and
the α cells to norepinephrine stimuli (25).   The hypoglycemia, the low insulin

Glucose (mg/dl)

Day of gestation	
18	35 ± 1
20	48 ± 4
21	48 ± 4
Time after birth	
0	5 ± 1
15 min	5 ± 1
30 min	5 ± 1
60 min	39 ± 2
90 min	40 ± 3
24 h	50 ± 3

Fig. 9: Foetal blood glucose concentration during the perinatal period.
Rat foetuses were delivered on the day indicated in the table (18-21)
by cesarean section, or born naturally after 21 days of pregnancy.
Each value represents the mean ($\pm$S.D.) of 5 independent experiments.

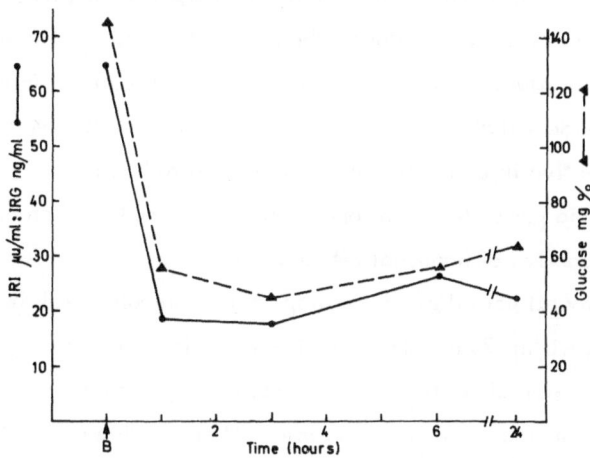

Fig. 10: Concentrations of the insulin glucagon ratio and glucose concen-
trations in the serum of human newborns.
IRI = Immunoreactive insulin; IRG = Immunoreactive glucagon
B = Birth
Adapted from F. Blazquez, 1974.

glucagon ratio, the rise in cyclic AMP, all contribute to a cessation of glycogen synthesis by converting glycogen synthetase a to b and the initiation of glycogenolysis by activating glycogen phosphorylase kinase and inactivating glycogen phosphorylase phosphatase. The role of glucagon in the induction of observed changes in hepatic metabolism is evident since the rise in endogenous glucagon precedes the glycogenolysis. Exogenous glucagon administered to fetal rats in utero has been shown to be capable of provoking premature glycogenolysis (31, 32). At the same time insulin, capable of antagonising glucagon effects dropped. In addition to the hormonal effect, a direct action of glucose on the glycogenolytic enzymes may be operative as well. Girard (25) showed that the prevention of hypoglycemia by glucose administration impaired phosphorylase activity. Glucose administration did not decrease glucagon secretion in the neonatal rat pancreas. On the other hand, the effect of glucose could not be easily explained by its ability to increase insulin secretion, since insulin had no effect on liver phosphorylase in the newborn rat. Furthermore in the isolated adult rat liver, glucose caused a rapid decrease of phosphorylase activity independent of changes in levels of cAMP, insulin or glucagon (33). This suggested that glucose can have a direct modulating effect on glycogen metabolism in the newborn rat as in the adult. Glucose is known to affect the affinity of phosphorylase phosphatase to phosphorylase a. Gross measured during the prenatal period the Ka of phosphorylase phosphatase for glucose and found it similar to the Ka known in adult rats (7.0 mM). It is evident that as glucose levels decrease to 0.35 mM this enzyme cannot be very effective in converting phosphorylase a to b (34).

The hypoglycemia per se has therefore a tendency to maintain an active phosphorylase irrespective of hormonal effects.

During the postnatal period glycogen drops from a prenatal peak level of 60 mg/g to 14 mg/g within 90 minutes. At the same time, pepcarboxy kinase is induced and gluconeogenesis initiated (25). From this period onwards hormonal controls of glycogen metabolism, based on intermittent feeding and fasting periods, are not much different from the adult pattern. The stormy perinatal period represents a major adaptation of the organism to a changing environment. This change requires a complex interaction of enzymatic, hormonal and neuroendocrine activities, which enable the newborn to switch from a more or less constant environment to a changing world. In case of disease states such as maternal diabetes, pre- or post-maturity or glycogen storage disease, the delicate balance between the various regulatory mechanisms in the perinatal period can easily be upset leading to severe, sometimes fatal sequelae.

REFERENCES

1. Devos, P, and HG Hers, Glycogen metabolism in the liver of the fetal rat. Biochem J 140: 331-340, 1974

2. Schwartz, AL, NCR Raiha, and TW Rall, Hormonal regulation of glycogen metabolism in human fetal liver. Diabetes 24: 1101-1112, 1975

3. Shelley, HJ, Glycogen reserves and their changes at birth and in anoxia. Br Med Bull 17: 137-143, 1961

4. Jost, A, The role of fetal hormones in prenatal development. Harvey Soc Lect 55: 201-226, 1961

5. Girard, JA, A Jost, and JP DuPouy, Modification precoces des hepatocytes de foetus de rat de capites sous l'influence du cortisol. CR Acad Sci 267: 857, 1968

6. Greengard, O, and HK Dewey, The premature deposition or lysis of glycogen in livers of fetal rats injected with hydrocortisone or glucagon. Develop Biol 21: 452-461, 1970

7. Pines, M, N Bashan, and SW Moses, Effect of hydrocortisone and glucagon on glycogen metabolism in the fetal rat liver. Biochim Biophys Acta 411: 369-375, 1975

8. Eisen, HJ, ID Goldfine and WH Glinsmann, Regulation of hepatic glycogen synthesis during fetal development: Roles of hydrocortisone, insulin, and insulin receptors. Proc Nat Acad Sci USA 70: 3454-3457, 1973

9. Vanstapel, F, F Dopere, and W Stalmans, Glucocorticoid-dependent development of glycogen synthase phosphatase in foetal rat liver. Abstracted in the control of glycogen metabolism, Brussels 5-7 Sept 1979 FEBS Adv Course No 65 Abstr No 55

10. Plas, C, F Chapeville, and RL Jacquot, Development of glycogen storage ability under cortisol control in primary cultures of rat foetal hepatocytes. Develop Biol 32: 82-91, 1973

11. Murphy, BEP, Steroid arteriovenous differences in umbilical cord plasma: Evidence of cortisol production by the human fetus early in gestation. J Clin Endocrinol Metab 36: 1037-1038, 1973

12. Plas, C, and J Nunez, Role of cortisol on the glycogenolytic effect of glucagon and on the glycogenic response to insulin in fetal hepatocyte culture. J Biol Chem 251: 1431-1437, 1976

13. Eisen, HJ, WH Glinsmann, and P Sherline, Effect of insulin on glycogen synthesis in fetal rat liver in organ culture. Endocrinology 92: 584-588, 1973

14. Plas, C, P Menuelle, MLJ Moncany, and C Fulchignoni-Lataud, Time Dependence of the glycogenic effect of insulin in cultured fetal hepatocytes. Diabetes 28: 705-712, 1979

15. Schwartz, AL, and TW Rall, Hormonal regulation of glycogen metabolism in neonatal rat liver. Biochem J 134: 985-993, 1973

16. De Wulf, H, and HG Hers, The role of glucose, glucagon and glucocorticoids in the regulation of liver glycogen synthesis. Eur J Biochem 6: 558-564, 1968

17. Miller, E, A Gerbretsen, C Clark, JF Ashmore, and DO Allen, Effect of glucagon and dibutyryl cyclic AMP on phosphorylase activity and gluconeogenesis in rat liver slices. Pro Soc Exp Biol Med 146:186-189, 1974

18. Schaub, J, and I Becker, The effect of glucagon on the development of glycogen phosphorylase and glycogen synthetase in rat liver. Biochem Biophys Acta 279: 398-400, 1972

19. Pines, M, Master thesis presented to the Department of Biology, Ben Gurion University of the Negev, Beersheba, 1975

20. Butcher, RW, and VR Potter, Control of the adenosine 3'5' monophosphate-adenyl cyclase system in the liver of developing rat. Cancer Research 32: 2141-2147, 1972

21. Novak, E, GI Drummond, J Skala, and P Hahn, Developmental changes in cyclic AMP, protein kinase, phosphorylase kinase and phosphorylase in liver, heart and skeletal muscle of the rat. Arch Biochem Biophys 150: 511-518, 1972

22. Assan, R, and J Bolloit, Pancreatic glucagon and glucagon-like material in tissues and plasmas from human fetuses 6-26 weeks old. Jonxis,JKP, HKA Visser, and J Troelstra, in Metabolic Processes in the fetus and newborn infant. Ed Baltimore, Williams and Wilkins 1971

23. Hommes, FA, and A Beere, The development of adenyl cyclase in rat liver, kidney, brain and skeletal muscle. Biochim Biophys Acta 237: 296-300, 1971

24. Girard, JR, and N. Zeghal, Adrenal catecholamine content in fetal and newborn rats. Biol Neon 26: 205-213, 1973

25. Girard, JR, D Caquet, and I Guillet, Control of rat liver phosphorylase, glucose-6-phosphatase and phosphoenolpyruvate carboxykinase activities by insulin and glucagon during the perinatal period. Enzyme 15: 272-285, 1973

26. Di Marco, PN, AV Ghisalberti, CE Martin, IT Oliver, Perinatal changes in liver corticosterone, serum insulin and plasma glucagon and corticosterone in the rat. Eur J Biochem 87: 243-247, 1978

27. Girard, JR, GS Cuendet, EB Marliss, A Kernran, M Rieutort, and R Assan, Fuels, hormones and liver metabolism at term and during the early postnatal period in the rat. J Clin Invest 52: 3190-3200, 1973

28. Bashan, N, Y Gross, S Moses, and A Gutman, Rat liver glycogen metabolism in the perinatal period. Biochim Biophys Acta 587: 145, 1979

29. Blazquez, E, T Sugase, M Blazquez, and PP Foa, Neonatal changes in the concentration of rat liver cyclic AMP and of serum glucose, free fatty acids, insulin, pancreatic and total glucagon in man and in the rat. J Lab Clin Med 83: 957-967, 1974

30. Girard, JR, R Assan, and A Jost, Glucagon in the rat fetus in fetal and Neonatal Physiology, Proceedings of the Sir J Barcroft Centenary Symposium, Cambridge University Press, London

31. Cornblath, M, ML Parker, SH Reisner, RE Forbes, and WH Daughaday, Secretion and metabolism of growth hormone in the premature and full term infants. J Clin Endocrinol Metab 25: 209-218, 1965

32. Girard, J, and D Bal, Effect du glucagon zinc sur la glycemie et le teneur en glycogene du foie fetal du rat en fin de gestation. CRH Acad Ser D 271: (P + 1) 777-779, 1970

33. Glinsmann, WH, G Pauk, and E Hern, Control of rat liver glycogen synthetase and phosphorylase activities by glucose. Biochem Biophys Res Commun 36: 931-936, 1970

34. Gross, Y, Perinatal glucose and glycogen metabolism.   Master thesis pre-
    sented to the Department of Biology, Ben Gurion University of the Negev,
    Beersheba, 1976.

EXTRAHEPATIC GLYCOGEN STORES
AND THEIR ROLE IN HOMEOSTASIS
R. De Meyer

INTRODUCTION

Since Claude Bernard (1), fetal glycogen accumulated in
late pregnancy has been considered as the major store for
energy supply during the early neonatal period. These stores were taught to be distributed in different organs
(lungs, muscle, liver) and the role of each tissue was supposed to contribute to maintain blood glucose at a normal
level. The role of the liver however gained so much importance in this regard that the role of the other tissues
became overlooked. Only a limited number of papers in the
li t erature of the last 20 years deal with extrahepatic
glycogen.
Lung glycogen has been studied by Sorokin (2) and Shelley
(3) and the role of this glycogen in differentiation of the
organs has been established by Sorokin (4).
Concentration of heart glycogen is high (5) and is related
to the resistance of the newborn to hypoxia as measured by
the time newborns can sustain oxygen withdrawal (6,7).
Brown fat has a high glycogen concentration for the purpose
of thermogenesis (8).
Carcass glycogen which is mostly located in muscles, can
not be utilized for restoring glucose level in blood for
muscle lacks glucose-6-phosphatase.However glycogen can be
mobilized and metabolized in loco and so a sparing effect
of blood glucose on blood is elicited.
The aim of this paper is to compare quantitatively the glycogen content of different organs in late pregnancy and

also to study the mobilization of these stores in the early
neonatal period. In a second step we intend to investigate
to what extent different factors (hormonal, environmental
and nutritional) can modify the normal evolution of these
stores.

1. Evaluation of glycogen stores during intrauterine life

It is a well known fact that during the end of gestation,
glycogen accumulates in the liver to very high concentrations (3, 9). Concomitantly, concentrations in lung and
heart tend to decrease whereas in skeletal muscle, glycogen
is still accumulating. These data have very well been
summarized by Shelley (3) but they have been gained from
different species and the information about the rat remains
sparse. Our personal data as it appears in Figure 1 may be
summarized as follows :

1) Liver glycogen is highly concentrated at the end of
   pregnancy. However the level at day 22 is lower than on
   day 21. This has been shown before by Brasseur et al.(9).
2) Lung glycogen decreases 5 times between the 20th and the
   21st day.
3) Heart glycogen is high in comparison to adult values.
4) Muscle glycogen expressed in terms of carcass glycogen
   is rather stable. It is possible that if the glycogen
   content of one specific muscle would have been investigated, differences could have been observed in relation
   to the differentiation in muscles.
5) Glycogen concentration in brown fat is high on day 20
   and decreases lightly on day 21.

Fig.1 GLYCOGEN CONCENTRATION IN DIFFERENT ORGANS IN LATE FOETAL
PERIOD IN THE RAT

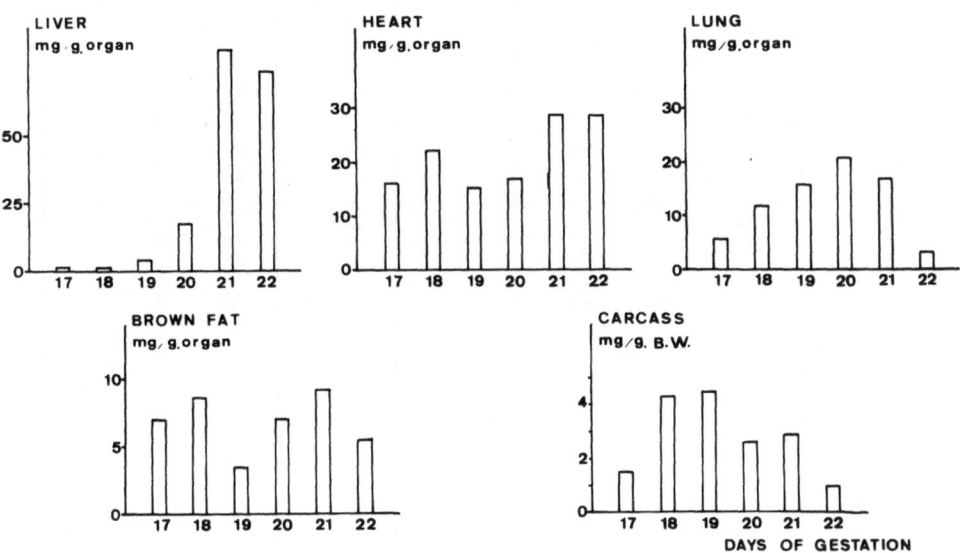

All these data are expressed in terms of concentration
(mg/gr/organ); from a cellular point of view this way of
expression fits very well; in respect to the general economy however the total amount of glycogen accumulated in each
organ is more representative. This may be expressed in mg
per gram body weight and represents the total contribution
of each organ to the carbohydrate pool.
This aspect is shown in Table 1.

Table I   CONTRIBUTION OF GLYCOGEN OF DIFFERENT ORGANS TO THE TOTAL BODY POOL. GLYCOGEN CONTENT IS EXPRESSED IN MG/GR BODY WEIGHT.

	17	18	19	20	21	22
Liver	0,149±0,056	0,144±0,067	0,370±0,237	1,34 ±0,47	5,77 ±1,62	3,99 ±0,64
Heart	0,160±0,021	0,147±0,041	0,122±0,042	0,122±0,03	0,170±0,050	0,180±0,045
Lung	0,176±0,060	0,35 ±0,16	0,66 ±0,21	0,78 ±0,10	0,64 ±0,23	0,098±0,044
Interscapular Brown fat	0,087±0,032	0,069±0,036	0,043±0,011	0,094±0,018	0,11 ±0,055	0,041±0,017
Carcass	1,56 ±1,47	4,32 ±0,88	4,49 ±1,25	2,60 ±1,50	3,27 ±1,33	1,03 ±0,56
TOTAL	2,13	5,03	5,69	4,94	9,96	5,34

From these data, it may be concluded that :
1. Although the amount of hepatic glycogen increases more
rapidly than the extrahepatic glycogen during the end of
gestation, the total amount of extrahepatic glycogen still
accounts for more than 50 %.
2. Among the extrahepatic glycogen stores, the most important are those accumulated in muscle. This large amount of
muscle glycogen plays probably an important role in glucose
homeostasis, although it is well known that muscle glycogen
can not be returned to blood glucose pool. However, it can
be utilized as fuel inside the cells and in this manner
has a sparing effect on the glucose pool in blood.
Next but far less in quantity are the stores in the lung
which decrease progressively towards the end of gestation.
Cardiac glycogen and glycogen of brown fat although highly
concentrated represent but a small amount in respect to the
total glycogen in the body.

2. Postnatal evolution of extrahepatic glycogen

After birth, liver glycogen starts disappearing after a lag
period of two hours (10, 8) and so too for the extrahepatic
glycogen stores. After a period of six hours about 5.5 mg
per gram body weight have been mobilized of which about 3 mg
per gram body weight originate from the extrahepatic stores;
what may not be neglected in balance studies.
Our data concerning the postnatal evolution of extrahepatic
glycogen are summarized in Table IIa and IIb. We did not
observe any difference in the pattern of postnatal evolution, whether the animals were prematurely delivered or at
term.

Table IIa   GLYCOGENOLYSIS IN THE POSTNATAL PERIOD IN THE RAT

		0'	60'	120'	180'	240'
Liver mg/g BW	premature	5,77± 1,62	4,95± 1,07	3,97± 0,87	3,12± 1,07	3,12± 0,85
	term	3,99± 0,64	3,56± 0,66	3,29± 0,50	2,29± 0,57	1,55± 0,72
Heart mg/g org.	premature	29,08±11,78	22,06± 5,92	26,65± 3,26	22,55± 7,52	29,16± 9,98
	term	28,83± 7,92	35,12± 7,34	28,10± 6,63	26,33± 4,50	21,76± 4,51
Lung mg/g org.	premature	16,97± 6,99	20,47±10,06	17,08± 4,57	17,58±18,39	16,90± 2,86
	term	3,27± 1,59	6,63± 3,35	5,38± 2,18	3,65± 3,26	1,29± 0,54
Brown fat mg/g org.	premature	9,19± 3,45	5,63± 1,72	3,73± 1,30	3,07± 1,20	2,86± 1,10
	term	5,50± 2,97	4,71± 2,38	3,06± 1,49	2,58± 1,95	1,43± 0,42
Carcass mg/g BW	premature	2,89± 1,51	4,79± 1,63	2,79± 2,53	2,29± 1,89	0,68± 0,14
	term	1,03± 0,56	1,05± 0,46	0,68± 0,17	0,84± 0,38	0,60± 0,16

Premature : 21st day of gestation

Term   : 22th day of gestation

Table IIb   NET LOSS OF GLYCOGEN DURING THE 240 FIRST MINUTES OF LIFE

	mg per g organ			mg per g B.W.		
	20 th day <4 g	20 th day >4 g	21 st day	20 th day <4 g	20 th day >4 g	21 st day
Liver	26,67	11,05	37,95	2,65	1,46	2,44
Heart	0	1,47	7,07	0,02	0	0,04
Lung	0,07	2,16	1,98	0,16	0	0,06
Brown fat	6,33	6,54	4,07	0,06	0,07	0,03
Carcass	-	-	-	2,21	2,41	0,43
Liver/carcass ratio				5,10 / 1,20	3,94 / 0,60	3,00 / 5,67

A remarkable fact is that shortly after birth, the concentration of glycogen in the organs increases. One may wonder whether this is due to a real glycogen synthesis or whether it is related to changes in the water content of these organs.

It is also noteworthy that the amount of glycogen disappearing from the stores even should it be entirely metabolized to CO_2 accounts but partly for the energy needs. Indeed, 5.5 mg carbohydrate will provide 22 cal/g/6 hours, whereas the total amount needed is calculated according to the diagram of Benedict (11) comes to 75 cal/g/6 hours.

The role of extrahepatic glycogen played in the neonatal period differs according to the organ considered.

For the lung, it is said that glycogen is essential for the cell division (2) and that glycogen disappears when the citric acid cycle activity increases (4).

The role of glycogen in the synthesis of surfactant is not clear yet (12). Supporting such a hypothesis, there is the fact that corticoids promote surfactant synthesis and decrease at the same time lung glycogen. However, during the first hour of life, when the need for surfactant is surely very high and its synthesis should be very active, no glycogen seems to be mobilized from the lung.

In prematurely born rats, this situation is different; glycogen concentration is high at birth (25 mg/g) but over a 6 hours period, glycogen has decreased by about 40 % and values are observed similar to those reached spontaneously in utero 12 hours later. One wonders whether in these prematures glycogen has been used to accelerate the maturation process (cell division - cell differentiation - morphologic transformation of the epithelium) or whether it has been used to increase the rate of surfactant synthesis.

Heart glycogen undergoes just like the other organs an increase of the concentration after 2 hours of extrauterine life and then returns to its normal condition. There is no net variation of glycogen concentration in the heart over a period of 6 hours. Concentration of glycogen in brown fat decreases also after birth, but this phenomenon is much more

pronounced in prematurely born than in term born animals.
This is perhaps related to the problem of thermoregulation
which is more acute in prematures.
Carcass glycogen decreases by about 50 % over a 6 hours period. The fate of this glycogen is actually not known but
plays probably an important role in muscular activity.

3. Alterations of extrahepatic glycogen stores by environ-
mental factors

1) Alterations of storage during intrauterine life
It is well known since the work of Jost (13), of Jacquot
(14) and of Jost and Picon (15) that liver glycogen storage
at the end of gestation is under control of adrenal cortex.
However little is known about the hormonal control of the
extrahepatic stores. Among the very rare data, a decrease
of concentration in heart glycogen has been described after
corticoids deprivation (16).
Our own investigation has been oriented along three lines;
we aimed
  a) to reproduce antenatally conditions that occur spon-
     taneously after birth. So for instance, increased
     oxygen tension.
  b) to investigate the effects of hormones known to be
     glycogenolytic such as adrenaline and glucagon.
  c) to evaluate the effect on extrahepatic glycogen of
     some drugs known to modify fetal liver glycogen
     concentrations.For instance barbiturates (Devos (17),
     De Meyer (18)),Insulin (De Meyer (19)).
Our data in respect of these three points may be summarized
as follows :
A. Hyperoxia. Maintaining the mother in a 100 % O_2 atmosphere during 24 or 36 hours in order to increase the arterial oxygen tension induces a slight hypoglycemia in the
fetus although the glycemia in the mother remains normal.
Glycogen stores in the liver, heart, lungs and brown fat are
not modified. In the carcass, glycogen concentration is increased by about 50 % (Table III).

Table III

EFFECT OF MATERNAL HYPEROXIA (24 h) ON GLYCOGEN STORES OF THE FOETUS 21 st DAY

	CONTROLS	TREATED	p
GLYCEMIA x	52,10+16,60 (6)	39,71+ 7,33 (11)	$>$0.05
LACTIC ACID x	95,74+21,62 (6)	111,79+28,61 (11)	n.s.
LIVER xx	5,47+ 1,62 (6)	4,32+ 1,05 (11)	n.s.
HEART xxx	29,08+11,18 (6)	23,95+ 2,63 (11)	n.s.
LUNG xxx	16,97+ 6,99 (6)	16,19+ 1,90 (11)	n.s.
BROWN FAT xxx	9,14+ 5,45 (6)	9,17+ 2,08 (11)	n.s.
CARCASS xx	3,27+ 1,33 (5)	5,97+ 0,72 (11)	$>$0.001

x mg/dl

xx mg/g B.W.

xxx mg/g organ

B. <u>Glycogenolytic drugs</u>. Before discussing the results, the question must be raised whether these drugs, given to the mother, reach the fetus or not. It is likely that adrenaline and reserpine do cross the placenta. However this is unlikely for glucagon (20). Therefore the possibility of an indirect action must be considered. It would of course be tempting to administer the drugs directly to the fetus to avoid the factor of placental permeability.

Such experiments have been performed under anesthesia (21, 22, 23). However as will be seen later, these experiments are to be considered very cautiously since anesthesia by itself modifies glycogen metabolism.

The effects of glycogenolytic drugs given to the mother in late gestation are summarized in Figure 2.

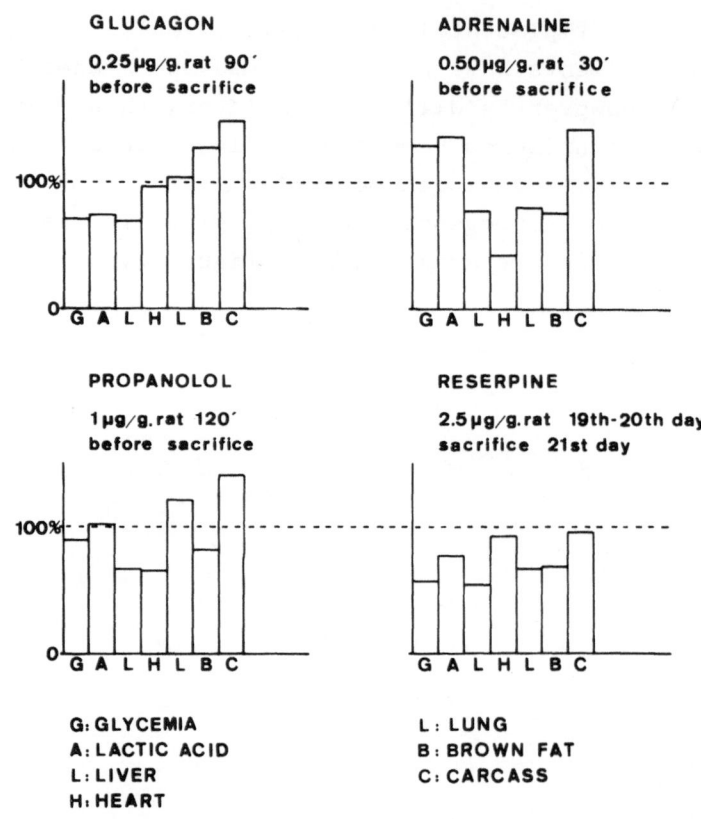

Fig. 2   EFFECT ON THE FOETUS OF DRUGS SUSCEPTIBLE
OF MODIFYING GLYCOGENOLYSIS AFTER
MATERNAL ADMINISTRATION

GLUCAGON

0.25 μg/g. rat 90′
before sacrifice

ADRENALINE

0.50 μg/g. rat 30′
before sacrifice

100% — — — — — — — — —

0 G A L H L B C

G A L H L B C

PROPANOLOL

1 μg/g. rat 120′
before sacrifice

RESERPINE

2.5 μg/g. rat 19th-20th day
sacrifice 21st day

100% — — — — — — — — —

0 G A L H L B C

G A L H L B C

G: GLYCEMIA
A: LACTIC ACID
L: LIVER
H: HEART

L: LUNG
B: BROWN FAT
C: CARCASS

Glucagon does not modify liver, heart, lung and brown fat
glycogen, but increases carcass glycogen. Adrenaline decreases heart and lung glycogen and increases carcass glycogen.
Reserpine, a drug known for its inhibitory activity on the
adrenal medulla produces a depletion of catecholamines and
decreases glycogen storage in the liver during the end of
gestation, but has no effect on heart and carcass glycogen.
However lung and brown fat are decreased.

C. Effect of other drugs. In Jost's experiments (13), corticoids have a replacement effect on liver glycogen synthesis
after adrenalectomy. Therefore we investigated the effect
of corticoids and of metopyrone on glycogen stores. The
results are mentioned in Table IV. Liver glycogen is not
changed. Glycogen storage is increased in heart, brown fat
and carcass, but is decreased in lung.

Table IV   EFFECT OF DEXAMETHASONE AND METOPYRONE

	GLYCEMIA mg/dl	LACTIC ACID mg/dl	LIVER mg/g B.W.	HEART mg/g org.	LUNG mg/g org.	BROWN FAT mg/g org.	CARCASS mg/g B.W.
Controls	45,85+16,61	105,24+24,05	4,48+0,96	24,24+3,82	24,23+4,90	14,61+ 3,83	5,00+2,43
Dexamethasone 5 mg/day 17→20th day	115,40+52,15	128,33+19,02	4,03+1,26	44,60+11,04	8,50+3,13	31,26+15,13	7,88+1,75
Metopyrone 30 mg/day 15→20th day	38,95+ 9,54	106,89+42,06	3,22+1,16	13,60+ 5,13	15,81+5,82	5,55+ 1,61	4,42+0,66

Barbiturates administered chronically from the 15 th to the
20 th day at a dose of 25 mg/kg has no effect on liver glycogen storage. Given acutely 180 minutes before sacrifice,
barbiturates appear highly glycogenolytic in liver, reducing the level to 25 % of the normal values, as has been
observed before yet by Devos and Hers (17). Barbiturates
as well chronically as acutely administered show no effect
on heart, lung and carcass glycogen but increase brown fat
glycogen.
Insulin (0.5 U/rat) administered 90 minutes before sacrifice exhibits a potent glycogenolytic effect in the liver,
but is without effect on the extrahepatic glycogen stores.
Starvation from the 17 th to the 20 th day reduces liver
and brown fat glycogen, but is effective on the other glycogen stores.
Salicylates have been said to inhibit the glycogen storage
in the fetal liver (24); these data could not be confirmed
in our investigation. After administration of sodium sali-
cylate (200 μg/g from the 17 th to the 19 th day) liver
glycogen was normal but its concentration in brown fat
fell to 50 % of the normal values.
After administration of triiodothyronine (17 μg/100 g)
from the 17 th to the 20 th day the glycogen stores are
very reduced in all organs, except carcass (Figure 3). This
is rather an exhaustion than a lack of synthesis for the
same lowering can be obtained by giving a single dose on
the 20 th day. The mechanism of T3 effect is not clearly
understood. Similar effects have been obtained with T3
analogs (25).

42

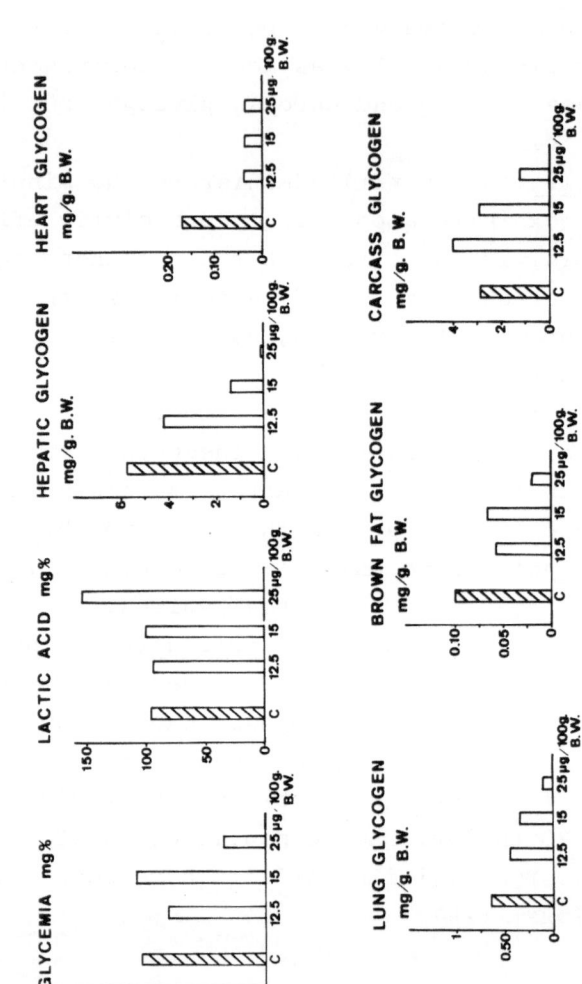

Fig.3 EFFECT OF T₃ ADMINISTERED TO THE MOTHER (17ᵗʰ TO 20ᵗʰ DAY)
ON CARBOHYDRATE METABOLISM OF NEWBORN RATS

C : Controls

From these data, it may be concluded that antenatally the
extrahepatic stores react entirely differently from the
liver. Generally speaking, extrahepatic stores seem more
stable than liver stores. Brown fat, however appears
more sensitive to several factors : a decrease is observed
after administration of reserpine, salicylates and after
starvation; a slight increase has been observed after barbiturates administration. The consequences of these changes
of glycogen stores on the postnatal adaptation to hypothermia remains to be investigated.

2) Changes in postnatal glycogen stores
In the early neonatal period, glycogen stores in different
organs evolve differently. Lung and heart glycogen remain
rather stable, brown fat and carcass on the contrary loose
part of their glycogen content. What are the factors regulating these particular evolutions ? Are the postnatal environmental changes responsible therefore ? With these questions in mind, several factors have been tested either with
the aim inhibiting glycogenolysis or with the aim inducing
it. Among the effects obtained, we note : lung glycogen
appears very stable and no factors tested have been able to
change the normal evolution. This observation is in favour
of the suggestion that lung glycogen has more to do with
differentiation processes than with supply of substrates
to the general economy. Concentration of glycogen in the
heart is said to be related to oxygen supply of the tissues;
an increased glycogenolysis is observed when the newborn
is maintained under hypoxic conditions leading to complete
exhaustion of the heart glycogen when the newborns are
maintained in a 100 % nitrogen and die (6). However if
pregnant rats (21 st day) are maintained for 15 minutes in
100 % nitrogen, their fetuses do not show any change in
their cardiac glycogen. Chronic hypoxia of the newborns
by maintaining them in a 10 % oxygen atmosphere for times
as long as 240 minutes, has no effect on cardiac glycogen,
although the lowering of oxygen tension is sufficient to
induce a hyperlactacidemia over 150 % of normal values.

It is also noteworthy that intrauterine life is normally
characterized by severe hypoxia and yet under these conditions, glycogen synthesis in the heart is not inhibited.
In vitro however, the conditions are very different and
heart muscle incubated under anaerobic conditions at 37°C
looses very quickly their glycogen whereas under incubation
in a oxygen rich atmosphere glycogen does not disappear.
It is also of interest to note that when newborns delivered
prematurely are kept in a 100 % oxygen atmosphere, some do
well and some are unable to adapt to extrauterine life.
These latter become cyanotic, breathe irregularly and are
rather hypotonic.
We investigated the glycogen stores of these animals and
compared with normal healthy animals of the same age.
Figure 4 shows the evolution of different parameters of
these newborns. The high values of lactic acid are the
expression of the serious hypoxic state. Nevertheless,
glycogen stores in liver, lung, brown fat and carcass
remain normal. Heart glycogen however is reduced by 50 % and
this since birth.

45

Fig.4 EVOLUTION OF DIFFERENT PARAMETERS IN NEWBORN RATS WITH
SPONTANEOUS INADAPTATION TO EXTRA-UTERINE LIFE

• Controls:in good health
▪ Poor adaptation(cyanosis irregular breathing)

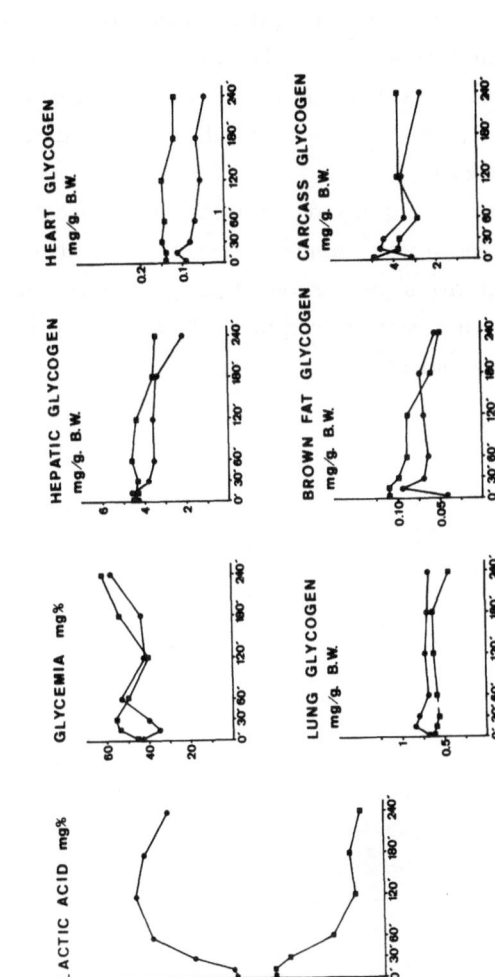

It has been taught since the investigation of Shelley (26)
that the exhaustion of heart glycogen in infants dying in
the early neonatal period is the consequence of anoxia.
In the light of our experiments, showing no effect when
energy supply is reduced artificially and on the other
hand showing exhaustion in naturally unwell newborn rats,
one would wonder whether - in vivo at least - the primary
event leading to death could not be the exhaustion of heart
glycogen for an unknown reason which in turn induces cardiac
failure and anoxia.
Adult heart muscle is said to utilize lactate preferentially to glucose. However administration of lactate in the
neonatal period does not have the glucose and glycogen
sparing effect one would expect. The essential data of this
experiment are summarized in Figure 5.

Fig. 5 EFFECT OF LACTATE ADMINISTRATION TO NEWBORN RATS

(50μl at 5% every 30´)

• Controls
■ Treated rats

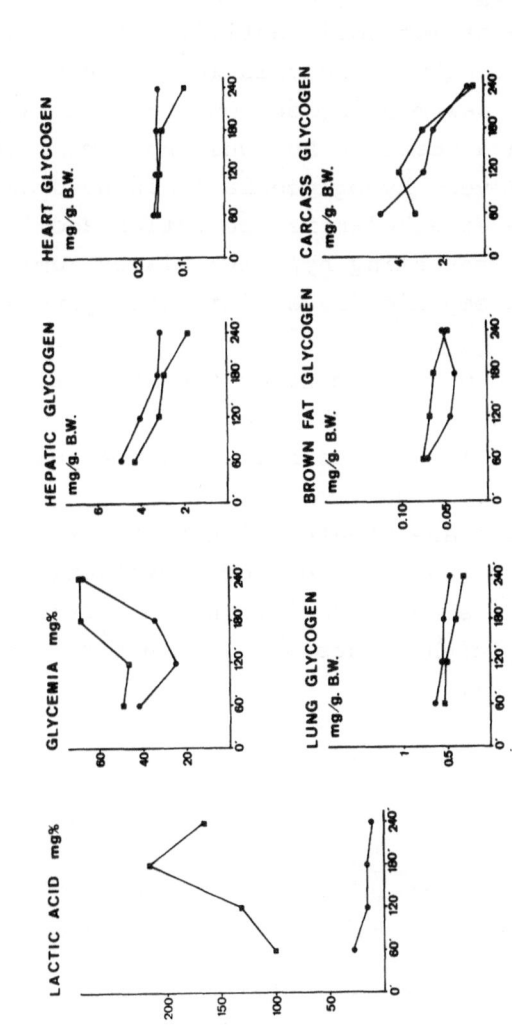

Brown fat is an important organ for thermoregulation in the
newborn. Glycogen is supposed to provide energy as well as
glycerol-1-phosphate for resynthesis of triglycerides, keeping the cycle for thermogenesis ongoing. One would then
expect that glycogenolysis in brown fat is directly related
to the intensity of metabolic activity of this organ. When
incubated at 38°C, glycogenolysis may be supposed to be
reduced when the newborn is kept on 32°C. In fact, we did
not observe such a response but when newborn rats are incubated at 38°C, severe hypoglycemia is induced and this may
partly stimulate catecholamines secretion and increase brown
fat metabolism. Preventing hypoglycemia by administration
of glucose is accompanied by a diminished glycogenolysis
in this fat.
Against this view we must mention the observation that
neither adrenaline nor blocking agents (α or β) given at
birth are able to modify the normal glycogenolysis in brown
fat. De Meyer (preliminary results).
When pregnant rats are treated with barbiturates (25 mg/kg)
from the 15 th to the 20 th day of gestation, brown fat
glycogen slightly elevated at birth, increases still more
and reaches very high values at 120 minutes (300 % of the
normal) (Figure 6).

Fig.6 EFFECT OF BARBITURATES ADMINISTERED
FROM THE 15th TO THE 20th DAY(25mg/kg/day)
ON BROWN FAT GLYCOGEN

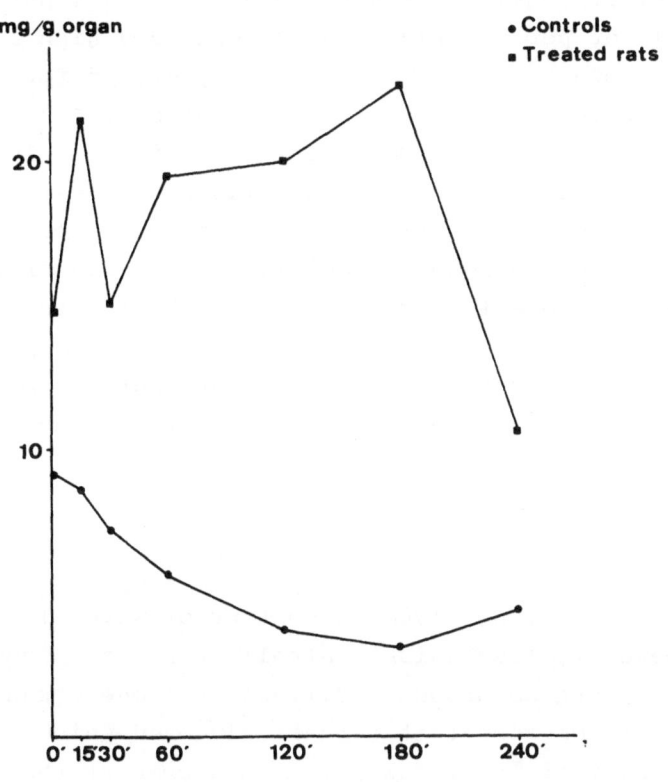

To a less extent, the same is observed in the liver after
birth. The reason for this glycogen accumulation remains
unexplained. The question one may ask is whether this accumulation is related to an increased glycogen synthesis -
in which case this may be a better adaptation to cold -
or whether it is due to a blocking effect on the glycogen
utilization - in which case it must be an impairment in the
adaptation to cold.

Carcass glycogen decreases in the neonatal period probably
since glycogen is utilized locally. When large amounts of
glucose (25 μl at 30 %) are given every 30 or 60 minutes,
subcutaneously or per os, glycemia is kept very high but
glycogen breakdown keeps on going normally except for carcass glycogen which reaches high concentrations.

These investigations on the postnatal evolution of glycogen
stores show once more the particularities of the extrahepatic stores. More especially we want to draw attention
on the brown fat which reacts specifically to some of the
drugs tested. The role this fact plays in the adaptation
to extrauterine life is very important and it may be suggested that unwanted effects of some drugs are perhaps mediated through the changes induced in brown fat glycogen.

CONCLUSION

At the end of gestation, glycogen content of several organs
is high. However the mechanism controlling glycogen homeostasis in these organs probably differs from one organ to
the other. It is obvious that carcass glycogen which represents nearly half of the total glycogen stores at that time
is not controlled by the same factors as liver glycogen.
After birth in several organs, glycogen is broken down but
the meaning of this event seems to be different from one
organ to the other.
- In liver, glycogenolysis is meant for keeping glucose
  blood level normal.
- In carcass glycogen, it is used in loco for metabolic

needs, what has a sparing effect on blood glucose.
- In brown fat, glycogen originating from glycogenolysis
  is partly used to forward energy for thermoregulation.
- In lung, changes are related to differentiation.
- In heart, changes of glycogen stores are not well under-
  stood and need further investigation.

REFERENCES

1. Bernard, C, De la matière glycogène considérée comme condition au développement de certains tissus chez le foetus, avant l'apparition de la fonction glycogénique du foie. C R Acad Sc 48:673-684, 1859.

2. Sorokin, S, HA Padykula and E Herman, Comparative histochemical patterns in developing mammalian lungs. Devel Biol 1:125, 1959.

3. Shelley, HJ, Glycogen reserves and their changes at birth. Brit Med Bull 17:137-143, 1961.

4. Sorokin, S, Study of development in organ cultures of mammalian lungs. Devel Biol 3:60, 1961.

5. Picon, L and J Bouhnik, Evolution de la teneur du coeur en glycogène au cours de la vie foetale et néonatale chez le rat. C R Soc Biol (Paris) 160:249-252, 1966.

6. Stafford, A and JAC Weatherall, The survival of young rats in nitrogen. J Physiol 153:457-472, 1960.

7. Mott, JC, Ability of young mammals to withstand total oxygen lack. Brit Med Bull 17:144-147, 1961.

8. De Meyer, R, P Gerard and G Verellen, Role of carbohydrates in energy balance of prematurely and term born rats. Pediat Res 13:1065-1071, 1979.

9. Brasseur, L, M Isaac-Mathy and R De Meyer, Etude comparative de la teneur en glycogène du foie foetal et du placenta chez la rate gravide. Ann Endocr 19:431-441, 1958.

10. Snell, K and DG Walker, Glucose metabolism in the newborn rat. Biochem J 132:739-752, 1973.

11. Benedikt, FG, Vital Energetics. Carnegie Instit Wash Report n° 503, 1938.

12. Maniscalo, WM, CM Wilson and I Gross, Influence of aminophyeline and cyclic AMP on glycogen metabolism in fetal rat lung in organ culture. Pediat Res 13:1319-1322, 1979.

14. Jacquot, R, Research on the endocrine control of glycogen accumulation in the liver of rat fetus. I. Experien-
ces of decapitation of the fetus in utero. J Physiol
(Paris) 51:655-721, 1959.

15. Jost, A and L Picon, Hormonal control of fetal development and metabolism, In : Advances in metabolic disorders,
Levine, R, and R Luft (eds.)4:123-184, 1970.

16. Picon, L and J Bouhnik, Action de la corticosurrénale
sur la teneur du coeur en glycogène chez le foetus de rat
et le rat nouveau-né. C R Soc Biol (Paris) 160:288-291,
1966.

17. Devos, P and HG Hers, Glycogen metabolism in the liver of
the foetal rat. Biochem J 140:331-340, 1974.

18. De Meyer, R, Effect of phenobarbital on carbohydrate
metabolism in the fetal rat liver. In Preparation.

19. De Meyer, R, Effect of insulin administered to the mother
on glycogen of the organs of the newborn rats. In Preparation.

20. Moore, WMO, BS Ward and C Gordon, Human placental transfer of glucagon. Clin Sci Molecular Med 46:125-129, 1974.

21. Hunter, DJS, Changes in blood glucose and liver carbohydrates after intrauterine injection of glucagon into
foetal rats. J Endocr 45:367-374, 1969.

22. Greengard, O and HK Dewey, The premature deposition or
lysis of glycogen in livers of fetal rats injected with
hydrocortisone or glucagon. Develop Biol 21:452-461,1970.

23. Girard, JR, D Caquet, D Bal and I Guillet, Control of
rat liver phosphorylase, glucose-6-phosphatase and
phosphoenolpyruvate carboxykinase activities by insulin
and glucagon during the perinatal period. Enzyme 15:
272-285, 1973.

24. Erikson, M, The effect of salicylates on the glycogen
content in the fetal liver and heart in two strains of
mice. Quoted by Larson, KS, Action of salicylates on
prenatal development p 171-186 in Malformations congéni-
tales chez les mammifères. Ed. Tuchmann-Duplessis.
Masson, 1971.

25. Kriz , BM, AL Jones and EC Jorgensen, Effects of a thyroid hormone analog on fetal rat hepatocyte ultrastructure and microsomal function. Endocrinology 102:712-722,
1978.

DISCUSSION

Qu. Minkowski : What is the role of the glycogen metabolism
in placenta in connection with glycogen in the fetus ?
An. De_Meyer_: Placental glycogen has not been considered
here. Although its role may be important in the intra-
uterine carbohydrate regulation, its role in the post-natal
adaptation seems rather limited.

Qu. Minkowski : The 30 % glucose solution used in your experiment seems too high because it might produce side
effects through hyperosmolarity ?
An. De_Meyer_: This is possible but if one would administer the same amount of glucose in more delute solutions,
the water intake would be too large.

Com. Shelley : I would like to support Professor De Meyer's
view of the importance of extra-hepatic glycogen stores.
These will be even more important in species, other than
the rat, which have very large stores of skeletal muscle
glycogen at birth. In the sheep, rhesus monkey and human
baby, the skeletal muscle glycogen concentration at birth
is at least five times the adult concentration, similar to
that in the adult liver, and in the pig it is ten times
the adult concentration.

FACTORS REGULATING BLOOD GLUCOSE IN THE FETUS AND IN THE NEWBORN

Colin T. Jones, Timothy Rolph, Geoffrey Band
and Enid Michael

The fetus receives a constant supply of glucose from the maternal circulation and this makes the major contribution to determining blood glucose concentration in the fetus (12). The gross effects of this are apparent when glucose transport across the placenta is limited by reduction of placental size (3,4) or by uterine artery ligation (5,6) where fetal blood glucose concentration is depressed. The extent of the depression of blood glucose is linearly related to the fetal growth rate (Fig. 1). Conversely infants of diabetic mothers probably have an increased rate of glucose transfer to the fetus and hence a higher fetal growth rate (7,8).

In the fetus blood glucose concentration regulates insulin secretion (9 - 11) although the sensitivity of the pancreas is less than in the adult (12,13). Also insulin determines fetal glucose consumption. This ensures that glucose supply to the fetus (14,15) determines the rate of glucose consumption (Fig. 1). Moreover an increase in fetal insulin concentration can increase the supply of glucose (16) to the fetus (Fig. 2).

Despite many conflicting reports (17 - 24) on the existence of gluconeogenesis in the fetus in vivo there is no convincing evidence that it occurs at a rate of quantitative importance. Indirect studies (25 - 27) while interesting do not provide conclusive proof and are open to

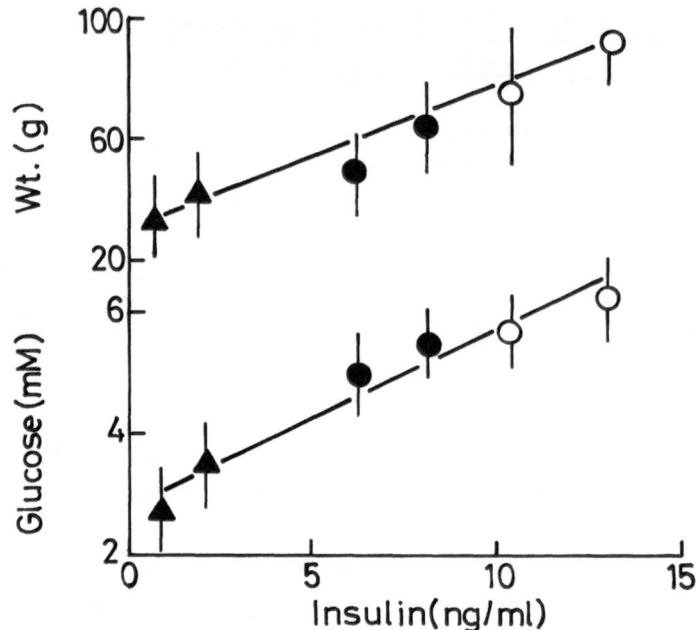

Figure 1: Relationship between blood glucose concentration, fetal growth and plasma insulin concentration in 60 - 63 days old fetal guinea-pigs.(O) controls, (●) naturally-occurring runts, (▲) growth-retarded fetuses after 30 days of uterine artery ligation. Results are means ± S.D. of 6 - 12 observations.

various interpretations. The major argument in favour of a gluconeogenic pathway in the fetal liver is the existence of all the gluconeogenic enzymes at relatively high activities for much of the latter half of gestation (28 - 34). Although phosphoenolpyruvate carboxykinase activity is probably exclusively mitochondrial. In the perfused liver of the near term fetus glucose synthesis from lactate, pyruvate or alanine does occur and is stimulated by glucagon (Fig. 3), but at the insulin/glucagon ratios in the fetus it is virtually inhibited (Fig. 3). Adrenaline also stimulates glucose synthesis in hepatocytes from the liver of the fetal sheep, monkey, guinea-pig and rabbit (35). An indication of the nature of the regulation of glucose synthesis in the fetal liver is seen in the perfused liver of 50-days old fetal guinea-pigs. At this time there

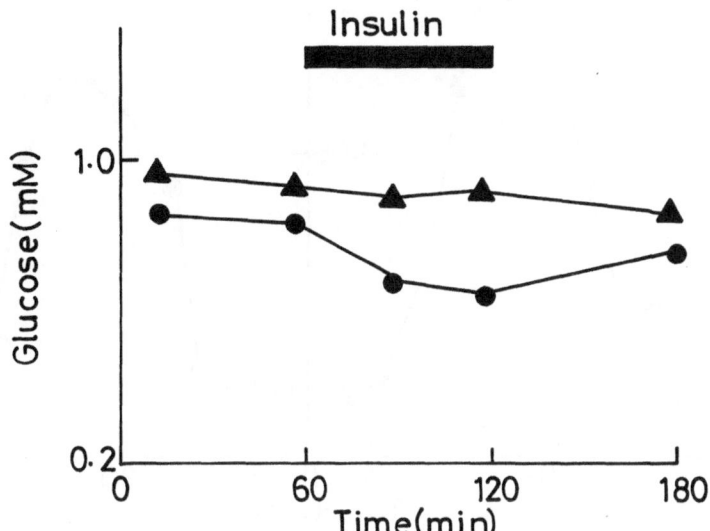

Figure 2: Changes in glucose concentration in the umbilical artery (—●—) and umbilical vein of a 129 days-old fetal sheep during the infusion of insulin (2 units/min) into the jugular vein.

is a full complement of gluconeogenic enzymes (30) but even in the presence of gluconeogenic precursors such as dihydroxy acetone,which substantially increases, the concentration of hexose phosphates in the fetal liver (Fig. 4a) there is no net glucose synthesis. On the addition of glucagon net glucose synthesis occurs (Fig. 4b) which cannot be accounted for by glycogen breakdown (glycogen content at this time is <0.2 mg/g of liver).

It is apparent that glucagon probably activates glucose 6-phosphatase or influences the accessibility of glucose 6-phosphate to it. In the absence of glucagon or in the presence of a high insulin/glucagon ratio glucose 6-phosphate cannot be converted to glucose. The gluconeogenic enzymes present in the fetal liver thus probably serve two major functions. They can participate in glycogen synthesis late in gestation (30). A more

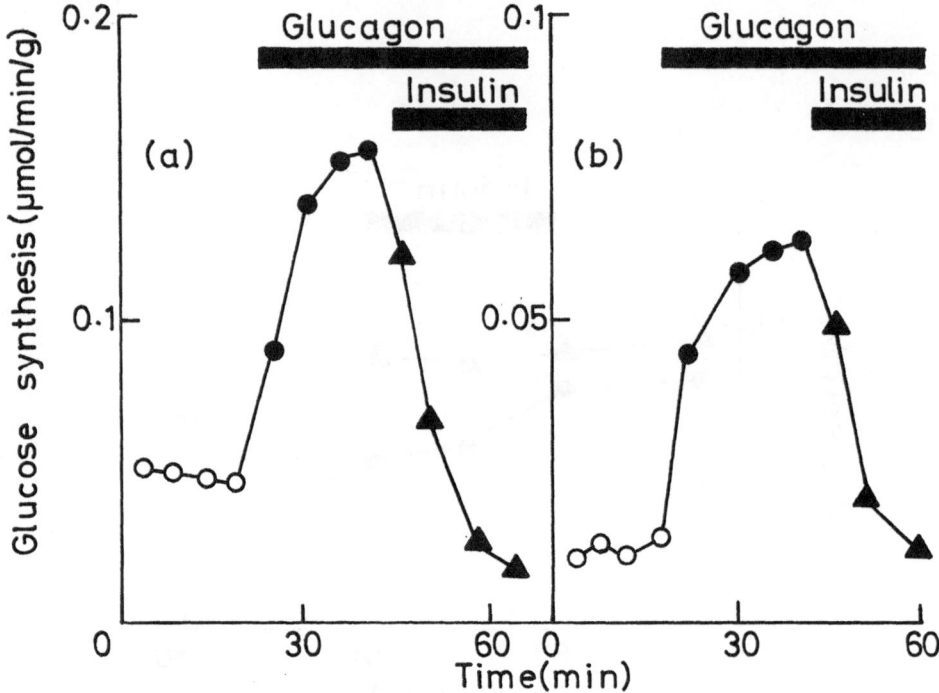

Figure 3: The incorporation of U-^{14}C alanine into glucose by the perfused liver of (a) 63-day fetal guinea-pig, (b) 31-day fetal rabbit. (-O-) control, (-●-) glucagon - 1 ng/ml (-▲-) glucagon - 1 ng/ml and (a) guinea-pig insulin - 50 ng/ml or (b) porcine insulin - 20 ng/ml.

important role is likely to be the maintenance of glucose 6-phosphate production to sustain the pentose phosphate pathway at times of high rates of fatty acid synthesis.

The role of the fetal liver in the autoregulation of blood glucose is likely to be minor by comparison with the adult liver. The fetal liver contains no glucokinase (37, 38) and glucose uptake is potently inhibited by alternative substrates like fatty acids, acetate, lactate, alanine and pyruvate (39) that can be present in relatively high concentrations in the fetal circulation (Fig. 5). During the latter part of gestation the fetal liver accumulates large quantities of glycogen (40). Although these are required postnatally mobilisation can occur in utero. The perfused liver of the rat (41) rabbit and guinea-pig (Fig. 6) produce glucose in response to adrenaline or noradrena-

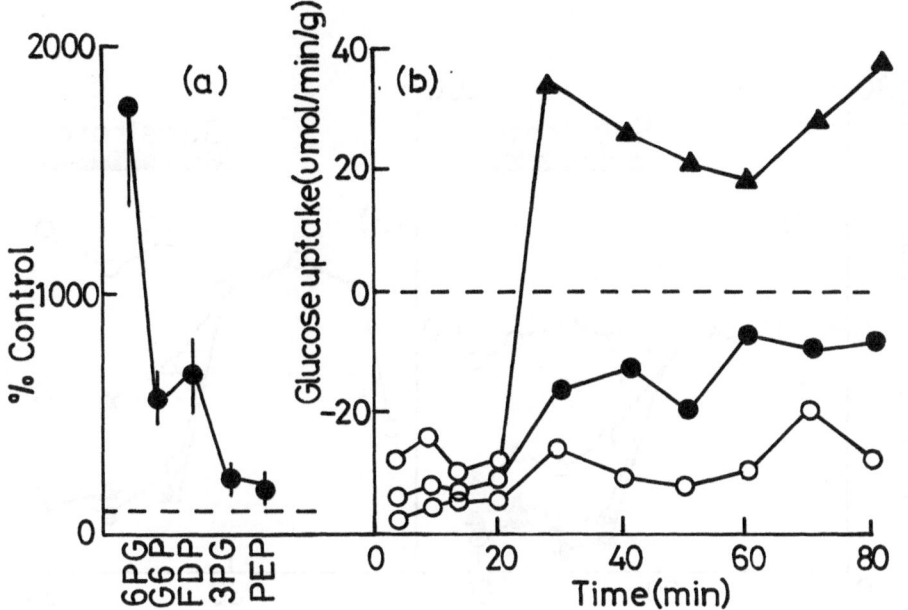

Figure 4: (a) Changes in metabolite concentrations
in the liver perfused with 10 mM-dihydroxyacetone
plus 0.1 mM-glucose as compared with 0.1 mM-glucose,
results are means ± S.D. of 5 experiments. (b) Glucose output from a liver perfused with (○) 0.1 mM-
glucose, (●) 0.1 mM-glucose plus glucagon - 1 ng/
ml or (▲) 0.1 mM-glucose, and 10 mM-dihydroxyacetone
plus glucagon - 1 ng/ml. Perfusion with 0.1 mM - glucose plus 10 mM-dihydroxyacetone was the same as
perfusion with 0.1 mM-glucose alone. Perfusions were
with 50 days-old fetal guinea-pig livers.

line stimulation. Physiological rates of adrenaline infusion into the fetal sheep elevate plasma glucose concentration (42) and cause differences in the umbilical arteriovenous difference of glucose that are consistent with
glycogen mobilisation in the liver (43). The major
mechanism, however, for the elevation of blood glucose is
the suppression of insulin secretion from fetal pancreas
by an α-adrenergic receptor mechanism (42,43). Although
hepatic glycogen is required to maintain blood glucose
after birth it is also likely to help sustain the supply of
glucose to the fetal heart and brain during labour,
particularly if severe fetal asphyxia is present (44). In
the fetal sheep labour causes, as in man (45), very high

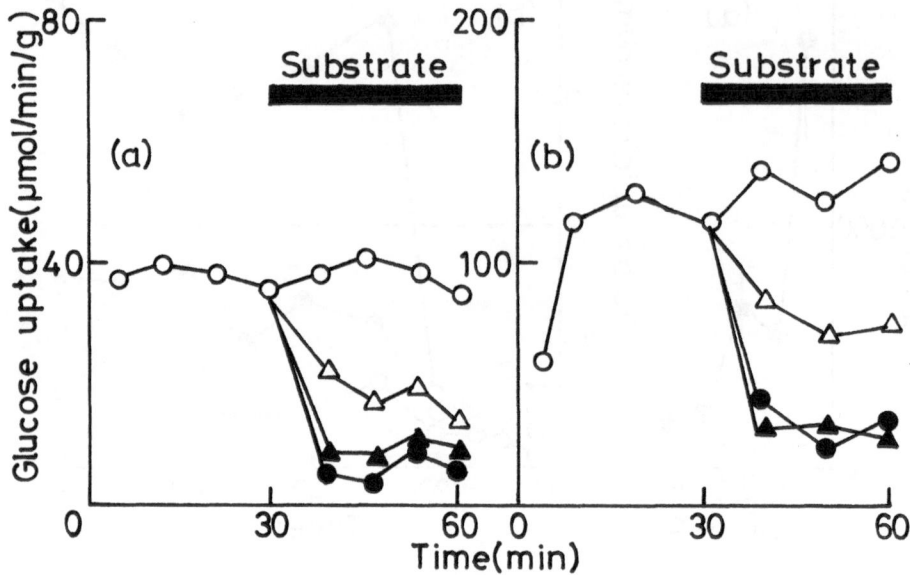

Figure 5: The inhibition of glucose uptake in the perfused liver of the 50 days-old fetal guinea-pig (a) or the 27-days old fetal rabbit (b). (—○—) control, (—●—) 0.5 mM-albumin bound palmitate, (—△—) 10 mM-lactate, (—▲—) 1 mM-acetate. Livers were perfused with Krebs' bicarbonate buffer containing 0.2 mM-glucose.

plasma adrenaline and noradrenaline concentrations that are probably responsible for the associated high plasma glucose concentration (Fig. 7).

In summary blood glucose in the fetus is controlled predominantly by the transfer of glucose across the placenta and secondarily by insulin and to a lesser extent catecholamines and glucagon. A high insulin/glucagon ratio is responsible for the inhibition of gluconeogenesis in the fetus.

Although most data is consistent with this view and a reduction in the insulin/glucagon ratio in the fetus caused for instance by maternal starvation may 'switch on' fetal gluconeogenesis (46), not all the data is consistent with this view. In growth-retarded fetus the insulin/glucagon ratio is much lower than normal (47,48) and can

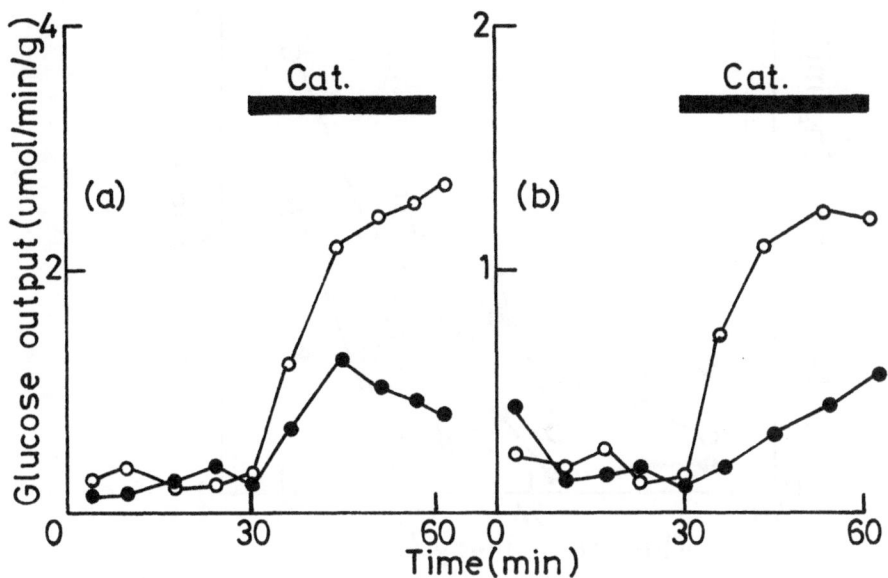

Figure 6: Glucose output from the liver of the 63 days-
old fetal guinea-pig (a) or the 31 days-old fetal rabbit
(b). (O●) control, (●) adrenaline 10 ng/ml, (O)
noradrenaline 10 ng/ml. Livers were perfused with Krebs'
bicarbonate buffer containing 1 mM-glucose.

be lower than in the newborn (Fig. 8). Despite this the
liver of these fetuses cannot synthesise glucose and has
low activities of the gluconeogenic enzymes (48). That
the liver of the growth-retarded fetal guinea-pig in vivo
is exposed to a high insulin/glucagon ratio is confirmed
by the demonstration that perfused livers of such fetuses
have a metabolite profile comparable to livers from
normal fetuses that have been perfused with glucagon
(Fig. 9). Another striking feature of the growth-retarded
fetus is that it does not necessarily have a low hepatic
glycogen concentration (48). In the growth-retarded
fetal guinea-pig for instance the glycogen concentration
is twice normal and fills most of the cytosol (48). The
mechanism of this effect is interesting as these fetuses
have blood glucose concentrations that are 40 - 60% of
normal and are markedly hypoinsulinaemic (48). Cortisol

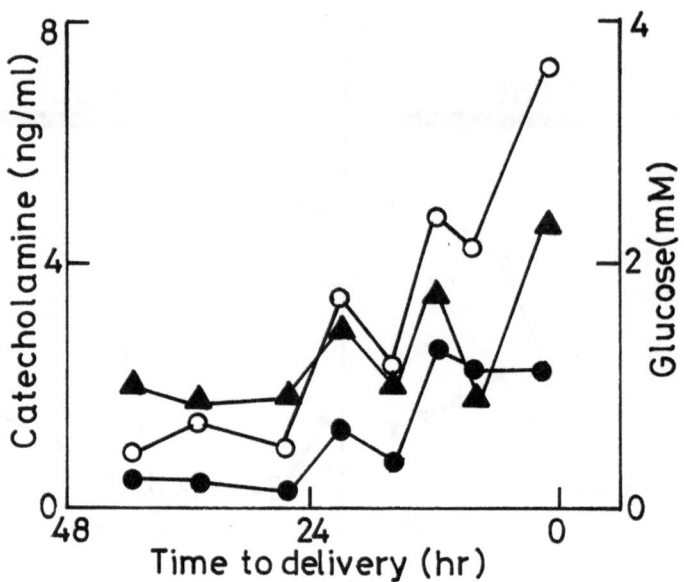

Figure 7: Plasma glucose (▲), adrenaline (O)
and noradrenaline (●) in fetal sheep during labour.
The ewe was 143 days pregnant.

concentrations are lower than normal although still relatively high and thyroid hormone concentrations are either
half normal or relatively unchanged (Fig. 10).

In the fetal rat glycogen deposition occurs at a time
of little glycogen turnover (49) and when glycogen synthetase activity is high   phosphorlylase a activity is
low and the rate of appearance of phosphorylase b kinase is
low (49-51). Furthermore glucagon administration to the
fetal rat does not elicit a large increase in hepatic
cyclic AMP (52) and in newborn rats it does not stimulate
glycogenolysis (53). This does not necessarily reflect the
situation in other species. Glucagon administration to
fetal sheep elevates blood glucose concentration and in
the perfused liver of the fetal rabbit or guinea pig it
causes large increases in cyclic AMP production and in
glycogenolysis (Table 1). In the fetal guinea pig glycogen
does turn over in vivo (Table 2) and there is phosphorylase

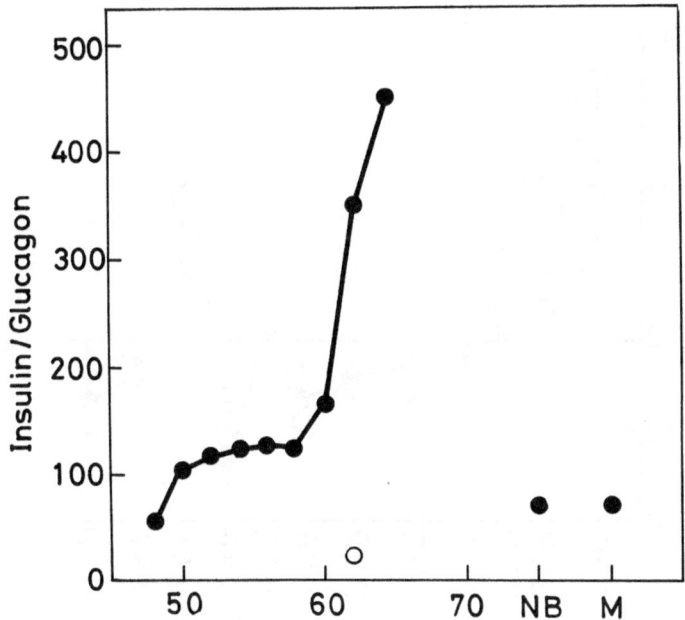

Figure 8: The insulin/glucagon ratio in the fetal
guinea-pig. (—●—) normal, (—○—) small fetus < 40%
of normal weight - caused by 30 days of uterine artery
ligation (—○—). The results are the means of 5 - 8
observations. N.B - 3 - 4 days old newborn, 1 -
2 days old newborn value is 49, M - maternal.

a activity that is relatively high at the time of glycogen
deposition (Table 3). Returning to the growth-retarded
fetal guinea pig phosphorylase activity is high. This
together with the low phosphofructokinase activity (48)
may account for the high rate of glycogen synthesis in the
perfused liver of the growth-retarded fetus and its higher
glycogen concentration. The cause of these changes, however, are not clear as the increase in the glucagon/insulin
ratio would be expected to have the opposite effect. This
suggests that other possibly unknown factors may be involved,
a point that will be persued below.

After birth glycogen synthetase a activity falls
dramatically and phosphorylase a activity rises sharply
over a period of 2 hours in the rat liver (51). This is
associated with a transient severe hypoglycaemia and a

64

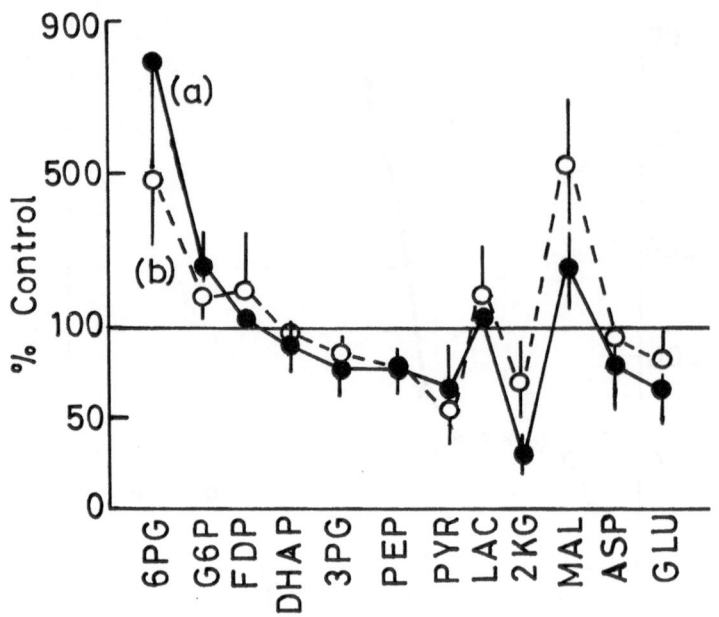

Figure 9: Changes in metabolite concentration in
the perfused liver of 49 - 51 days-old fetal guinea-
pig. (a) Liver of normal fetus perfused with medium
containing glucagon (1 ng/ml), (b) liver of growth-
retarded fetuses (< 40% of normal weight) perfused
with medium not containing glucagon. All livers
were perfused with Krebs' bicarbonate buffer containing 0.2 mM-glucose and the above results are for
comparison against the liver of a normal fetus perfused with medium alone. Results are means ± S.D.
of 4 - 6 experiments.

rapid loss of hepatic glycogen (51,54,55). At the same
time there is a sharp increase in the activity of the
gluconeogenic enzymes, particularly phosphoenolpyruvate
carboxykinase (56-58). These changes have been ascribed
to a sharp increase in the glucagon/insulin ratio and a
high plasma catecholamine concentration (59). Large
postnatal increases in the plasma glucagon/insulin ratio
have been reported associated with hypoglycaemia or
delivery in several species (60-62). Moreover, although
the changes are less pronounced than in the rat, there is
an increase in the activity of the gluconeogenic enzymes
and a depletion of glycogen (29-31, 63,64). The glycogen

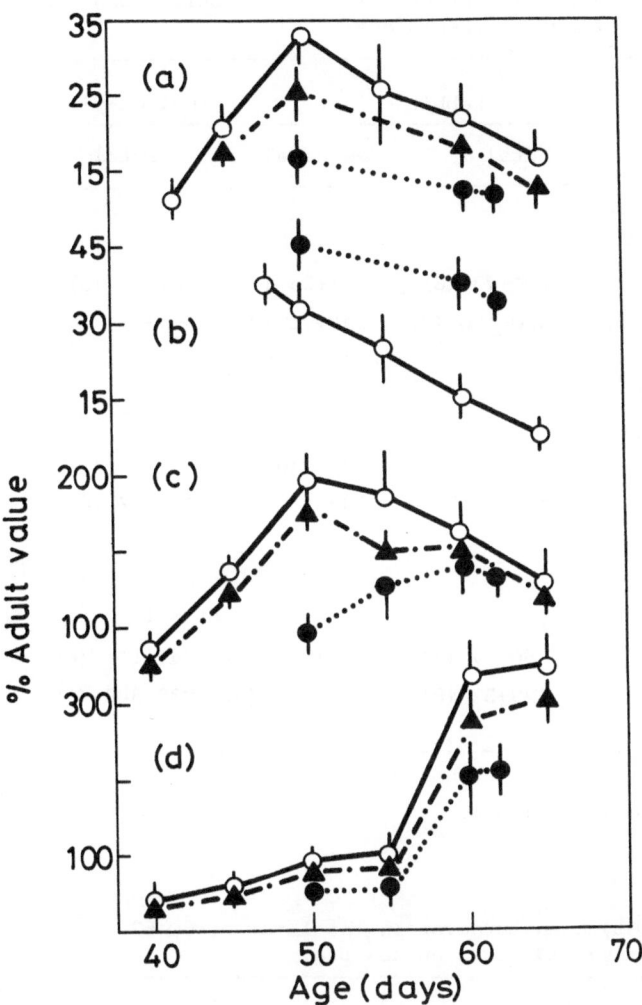

Figure 10: The relative changes in the concentration of plasma hormones in the developing guinea-pig fetus. (O) normal weight, (●) small fetuses after uterine artery ligation - 25-40% normal weight, (▲) naturally occurring small fetuses - 50-70% of normal weight. (a) guinea-pig insulin - assayed for this species specific hormone, (b) glucagon, (c) T_4, (d) cortisol. The results are means ± S.D. of 4 to 6 observations.

Table 1: Lactate, glucose and cyclic AMP production by
perfused livers from 26 days-old fetal and
6 days-old neonatal rabbits and 60-63 days-old
fetal guinea-pigs. Livers were perfused for
40 mins with Krebs' bicarbonate buffer containing 0.5 mM-glucose.

| | Rabbit | | Guinea-pig |
	Fetal	Neonatal	Fetal
Lactate production (n mol/min per g)			
Control	200+73 (6)	101+34 (7)	87+24 (6)
Glucagon (1 ng/ml)	510+150 (8)	140+51 (8)	398+114 (8)
cAMP production (p mol/min per g)			
Control	14+6 (5)	32+8 (6)	21+10 (7)
Glucagon (1 ng/ml)	89+20 (6)	194+73 (6)	183+52 (8)
Glucose production (n mol/min per g)			
Control	-50+27 (10)	152+84 (10)	347+101 (6)
Glucagon	129+52 (8)	962+327 (8)	937+348 (6)

Results are means + S.D.

Table 2: Glycogen turnover in the liver of 60-62
days-old fetal guinea-pig.

Time (hr)	1	4	8	16
Glycogen (mg/g)	15+6 (4)	18+5 (4)	14+7 (4)	17+8 (4)
^{3}H-glucose in glycogen (dpm/g of liver)	574+125 (4)	329+143 (4)	270+87 (4)	172+69 (4)

50 μ Ci of 6-^{3}H glucose was injected into the fetal guinea-
pig and fetuses were obtained at the times shown above.
Results are means + S.D.

Table 3: The activities (units/g) of enzymes of glycogen
synthesis in the liver of the fetal guinea-pig.

	Normal		Growth-retarded	
	49-51 days	60-63 days	49-51 days	60-63 days
Glycogen Synthetase a	0.006±0.003(4)	0.15±0.06(4)	0.007±0.004(4)	0.3±0.1(4)
Phosphorylase a	3.7±1.6(4)	5.3±2.5(4)	2.1±0.7(4)	2.3±0.9(5)
Synthetase kinase	0.07±0.03(4)	0.3±0.11(4)	0.09±0.04(4)	0.4±0.2(4)
Synthetase phosphatase	0.06±0.03(4)	0.72±0.23(4)	0.07±0.04(4)	0.79±0.4(4)
Phosphorylase kinase	0.09±0.03(4)	0.29_0.1(4)	0.05±0.02(4)	0.18±0.05(4)
Phosphorylase phosphatase	0.17±0.1(4)	0.96±0.24(4)	0.1±0.03(4)	0.41±0.18(4)

Enzyme activities were assayed essentially as outlined in (98).
Results are means ± S.D.

is thought to maintain blood glucose while hepatic gluconeogenesis is developing   .   Thus it is not unreason-
able to hypothesise that the major changes in hepatic
glycogenolysis and gluconeogenesis are caused by the increase in the glucagon/insulin ratio and by the secretion
of catecholamines.

As reported above in many species there is a full
complement of gluconeogenic enzymes in the liver at
delivery, except that phosphoenolpyruvate carboxylase is
largely mitochondrial. Thus in those species it has been
suggested that the changes in the hormonal environment
immediately after birth switches on this pathway. Furthermore the increased oxygenation of the liver after birth
has been proposed, by increasing mitochondrial and cytosolic
redox, to activate gluconeogenesis (65,66). However, the
redox of the fetal liver is not necessarily lower than that
of the adult (67,68). It is likely that the nature of the
changes in hepatic metabolism after birth vary substantially
across the species. While it is clear there are large
changes in the rat and possibly the rabbit (Table 4) the

metabolite profile of the rhesus monkey after birth is
virtually indistinguishable from that of the fetal rhesus
monkey (Table 4).

The extent to which our picture of the regulation of
the hepatic changes after birth is inadequate is further
demonstrated by a closer look at the growth-retarded newborn.

Table 4: Metabolite concentrations in freeze-clamped livers
from rabbits and rhesus monkeys.

	Metabolite concentration (n mol/g)			
	Rabbit		Rhesus monkey	
	29 days fetal	6 days neonatal	160-164 days-fetal	2-3 hr. neonatal
6-phospho-gluconate	3.9+1.0	0.5+0.3	2.0+1.3	1.5+0.7
Glucose 6-phosphate	80+18	61+21	142+87	103+29
Fructose 1,6-diphosphate	3.8+2.0	1.5+0.9	5.7+3.8	3.1+2.6
Glycerate 3-phosphate	81+27	158+38	41+20	70+30
Phosphoenol-pyruvate	27+12	72+24	24+13	**43+18**
Pyruvate	180+82	40+18	73+29	110+34
Lactate	3309+890	791+267	2390+987	2104+960
2-Ketoglutarate	87+30	32+17	64+31	162+60
Malate	394+140	68+13	229+104	179+84
Glutamate	2600+852	480+190	2312+610	2615+1009

Results are means ± S.D. of 4-6 experiments.

These have a high incidence of hypoglycaemia (69-73)
although insulin, glucagon and cortisol secretion (72-74)
and substrate supply (72,73) do not appear necessarily to
be impaired. It has been hypothesised that the pathway for
hepatic gluconeogenesis is abnormal in such infants and
increase in substrate supply such as alanine does not

elevate blood glucose concentration as it does in normal infants (75). However the cause of the hypoglycaemia has not been directly demonstrated although an impairment of glucose production has been shown in liver slices (a poor experimental technique) from growth-retarded newborn rats (76). It was reported above that the insulin/glucagon ratio in growth-retarded fetal guinea-pigs is lower than in normal fetuses and it seems likely that in the growth-retarded newborn it is not higher than normal. Despite this the gluconeogenic enzyme activities are low in the liver of the growth-retarded fetal guinea-pig and this condition, particularly for pyruvate carboxylase, persists for at least the first two days after birth (Table 5). Thus these newborns are hypoglycaemic and they have poor hepatic glycogen stores (Table 6). Moreover the fetuses with the lowest pyruvate carboxylase activity have the lowest plasma glucose concentration and the lowest hepatic glycogen concentration (Fig. 11). Not only is hepatic gluconeogenesis impaired but, contrary to the impression given by plasma alanine concentrations, the capacity for substrate supply, as indicated by muscle alanine aminotransferase, is also reduced (Table 7). This is probably related to the poor development of skeletal muscle in the growth-retarded newborn guinea-pig (Fig. 12). Relatively high ketone body concentrations in the plasma of small and normal newborn guinea-pigs at present suggest that fatty acid supply is not impaired in the growth-retarded newborn. The maintenance of a supply of free fatty acids has been implicated as an important point of control of glucose synthesis in the liver of the newborn (77,78) although it must be remembered that fatty acids do not stimulate glucose synthesis in all species (79,80).

Such data suggest that the changes in the insulin/ glucagon ratio after birth do not alone account for activation of glucose production and other factors are probably involved. Catecholamines have already been mentioned, they stimulate glycogenolysis and

Table 5: Gluconeogenic enzyme activities in the liver of
40-48 hours-old normal and growth-retarded new
born guinea-pigs.

	Enzyme activity (units/g)	
	Control	Retarded
Weight (g)	98.3+10.8	46.9+8.8
Pyruvate carboxylase	0.62+0.2	0.13+0.06 (22.4+8.1)
Phosphoenolpyruvate carboxy-lase	3.58+0.2	1.95+0.6 (55+12.9)
Alanine aminotransferase	1.49+0.80	1.54+0.52 (121+53)

Growth retardation was caused by uterine artery ligation at
day 30 of pregnancy.
Results are means $\pm$ S.D. of 7 experiments. Figures in
parentheses are % control values for paired newborn.
Enzyme activities are total tissue values assayed as
previously described (99).

Table 6: Plasma glucose and hepatic glycogen concentrations
in 40-48 hours-old normal and growth-retarded
newborn guinea-pigs.

	Control	Retarded	Control	Retarded
Weight (g)	99.7+6.9	43.6+8.1	101+12.7	59+4.8
Glucose (μmol/ml)	5.4+1.1	2.9 7+1.7	5.32+0.47	4.53+0.7
% control		(57+32)		(85+12.5)
Glycogen (μmol/g)	37.9+4.9	5.5+6.4	54.8+50.5	49.9+27.3
(μmol/ g liver)	145+202	9.3+11.4	198+160	106+58
(μmol/g of body weight)	1.5+2.1	0.2+0.2	2.0+1.9	1.79+1.0
n	5	5	6	6

Growth retardation was caused by uterine artery ligation at
day 30 of pregancy.

gluconeogenesis before and after birth in the liver of
several species (81-84). However the adrenal of the
growth-retarded fetal guinea-pig has high concentrations
of adrenaline and of noradrenaline. A group of hormones
that may have important effects are the gastrointestinal
polypeptides and pancreatic polypeptides. They can show

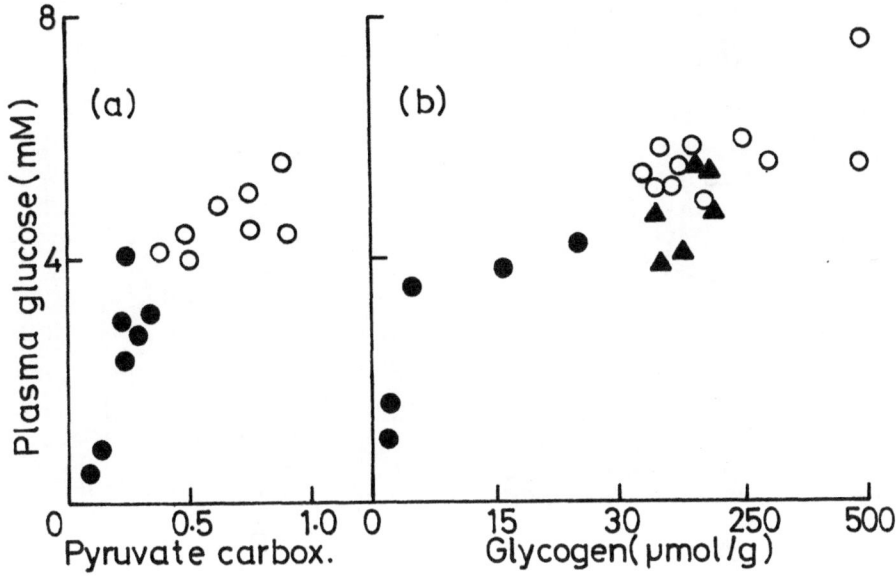

Figure 11: The relationship between plasma glucose
and (a) hepatic pyruvate carboxylase activity (units/
g) and (b) hepatic glycogen concentration (μmol glucose/
g). The measurements were made on 40-48 hours-old newborn guinea-pigs.   Normal guinea-pigs (100-120 g),
(O); small guinea-pigs <50 g, (●) and >50 g, (▲).
Growth-retardation was caused by uterine artery ligation
at day 30 of pregancy.

relatively large changes in plasma concentration after
birth (85,86) and pancreatic polypeptide for example is
likely to have large effects on carbohydrate metabolism
(87).

In addition to the changes in the rates of glucose
production in the liver of the newborn, its role in
autoregulation of blood glucose is quantitatively different from that in the adult (88,89). The glucose output
from the liver of the newborn is less sensitive to inhibition by glucose and insulin (89,90). This is likely
to be the result of little or no glucokinase activity (91,
92), high glucose 6-phosphatase activity (93-95) and
relatively low glycogen synthetase phosphatase activity(96).

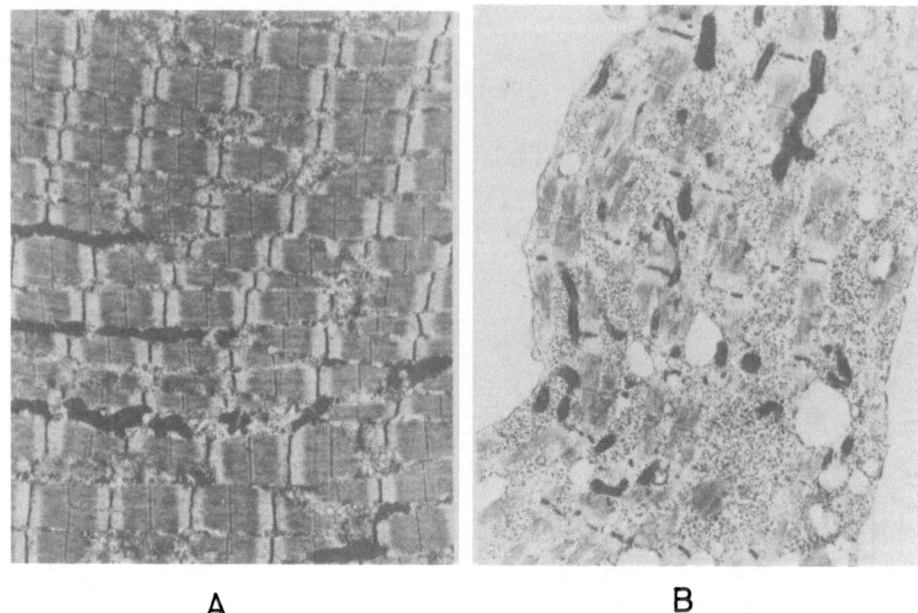

A   B

Figure 12: Electron micrographs of anterior hind-
limb muscle from 2 days-old newborn guinea-pig.
A - 100 g normal guinea-pig, B - 30 g growth-retarded
guinea-pig. Linear magnification is x 6500.

Table 7: Alanine aminotransferase in anterior hind-
limb skeletal muscle of 40-48 hours-old
normal and growth-retarded guinea-pigs.

| | Activity (units/g) | | |
	Control	Retarded	
Weight (g)	98.3+10.8	53.7+3.5	41.9+8.2
Alanine aminotransferase	3.86+1.23	3.56+0.3 (89+23.9)	1.48+0.28 (43.2+10.9)
n	7	3	4

Growth retardation was caused by uterine artery ligation
at day 30 of pregnancy. Figures in parentheses are %
control values for paired newborn. Enzyme activities are
total tissue values assayed as previously described (99).

The effects of some of these enzyme changes are clearly
seen in the responses of the perfused rabbit liver to
glucagon (Tables 1 & 8).

Table 8:   Metabolite concentrations in the perfused liver of
of the fetal and neonatal rabbit.

	Metabolite concentration (n mol/g)			
	29 days-old fetal		6 days-old neonatal	
	Control	Glucagon	Control	Glucagon
6-Phosphogluconate	2+1.0	70+19	1+0.5	0.9+0.4
Glucose 6-phosphate	61+12	310+82	19+7	72+20
Fructose 1,6-diphosphate	10+6	40+18	12+8.2	10.1+6.8
Phosphoenolpyryvate	25+11	29+12	15+7	34+19
2-Ketoglutarate	20+13	51+10	42+15	25+12
Malate	170+90	138+91	40+19	108+40

Results are means + S.D. of 6 experiments of livers perfused
with Krebs' bicarbonate buffer containing 0.5 mM-glucose with
or without glucagon (1 ng/ml).

In summary the role of the fetal and neonatal liver
in the regulation of blood glucose has been discussed.
The fetal liver can contribute to blood glucose through
glycogenolysis but it is unlikely to do so through
gluconeogenesis except possibly during maternal starvation. After birth hepatic glycogenolysis and gluconeogenesis are stimulated. This has been ascribed to the
rise in the glucagon/insulin ratio and although in the
newborn inhibition of insulin and glucagon secretion by
somatostatin will depress blood glucose (97), changes in
plasma insulin and glucagon or catecholamine do not appear
to alone explain the changes after birth.

Regulation of glucose consumption and insulin and
glucagon secretion are discussed elsewhere in this
symposium.

This work was supported by The Medical Research Council, The Royal
Society and Action Research for the Crippled Child. G.B. and E.M.
are MRC Scholars.

REFERENCES

1. Widdas, W. F. (1961) British Medical Bulletin, 17, 107.

2. James, E. J., Raye, J. R., Grosham, E. L., Makowski, E. L., Meschia, G. & Battaglia, F. C. (1972) Pediatrics, 50, 361.

3. Jones, C. T. & Robinson, J. S. (1979) In Maternal Effects in Development. Newth, D. R. & Balls, M. (Eds.), pp 395-409, Cambridge University Press.

4. Robinson, J. S., Kingston, J., Jones, C. T. & Thorburn, G. D. (1979) Journal of Developmental Physiology, 1 (in press).

5. Roux, J. M., Tardet-Coridroit, C. & Chanez, C. (1976) Biology of the Neonate, 15, 342.

6. Lafeber, H. N., Jones, C. T. & Rolph, T. P. (1979). In Nutrition and Metabolism of the Fetus and Infant. Visser, H. K. A. (ed.), pp 43-62, Martinus Nijhoff Publishers b.v., The Hague.

7. Obenshain, S. S., Adam, P. A. J., King, K. C., Raiha, N., Raivio, K., Teramo, K. & Schwartz, R. (1970) New England Journal of Medicine, 283, 566.

8. Persson, B. (1974) In Size at Birth. CIBA Foundation Symposium 27. pp 247-267, Associated Scientific Publishers, Amsterdam.

9. Bassett, J. M. & Thorburn, G. D. (1971) Journal of Endocrinology 50, 59.

10. Girard, J. R. (1974) Diabetes, 23, 310.

11. Fiser, R. H., Erenberg, A., Sperling, M. A., Oh, W. & Fisher, D. A. (1974) Pediatric Research, 8, 951.

12. Oakley, N. W., Beard, R. W. & Turner, R. C. (1972) British Medical Journal, i, 466.

13. Asplund, K., Westman, S. & Hellerstrom, C. (1969) Diabetologia, 5, 260.

14. James, E. J. Raye, J. R., Gresham, E. L., Makowski, E. L., Meschia, G. & Battaglia, F. C. (1972) Pediatrics, 50, 361.

15. Bassett, J. M. & Madill, D. (1974) Journal of Endocrinology, 62, 299.

16. Colwill, J. R., Davis, J. R., Meschia, G., Makowski, E. L., Beck, P. & Battaglia, F. C. (1970) Endocrinology, 87, 710.

17. Battaglia, F. C. & Meschia, G. (1978) Physiological Reviews, 58, 499.

18. Yeung, D. & Oliver, I. T. (1967) Biochemical Journal, 105, 1229

19. Ballard, F. J. & Phillipides, H. (1971) In Regulation of Gluconeogenesis. Soling, H. D. & Willms, B. (eds.), pp 66-81. Academic Press, New York & London.

20. Arinze, I. J. (1975) Biochemical and Biophysical Research Communications, 65, 184.

21. Warnes, D. M., Seamark, R. F. & Ballard, F. J. (1977) Biochemical Journal, 162, 627.

22. Prior, R. L. & Christensen, R. K. (1977) American Journal of Physiology, 233, E462.

23. Savon, P. M. J., Jeacock, M. K. & Shepard, D. A. L. (1976) Proceedings of the Nutrition Society, 35, 30A.

24. Prior, R. L. & Scott, R. A.(1977) Developmental Biology, 58, 384.

25. Goodner, C. J. & Thompson, D. (1967) Pediatric Research, 1, 443.

26. Morriss, F. H., Boyd, R. D. H., Makowski, E. L., Meschia, G., & Battaglia, F. C. (1973) Pediatric Research, 7, 794.

27. Hodgson, J. E. & Mellor, D. J. (1977) Proceedings of the Nutrition Society, 36, 33.

28. Aurricchio, S. & Rigillo, N. (1960) Biology of the Neonate, 2, 146.

29. Mersmann, H. J. (1971) American Journal of Physiology, 220, 1297.

30. Jones, C. T. & Ashton, I. K. (1976) Archives of Biochemistry and Biophysics, 174, 506.

31. Dawkins, M. J. R. (1966) British Medical Bulletin, 22, 27.

32. Raiha, N. C. R. & Lindros, K. O. (1969) Annales Medicinae Experimentalis et Biologiae Fenniae, 47, 146.

33. Ballard, F. J. & Oliver, I. T. (1965) Biochemical Journal, 95, 191.

34. Kirby, L. & Hahn, P. (1973) Pediatric Research, 7, 75.

35. Jones, C. T. (1980) In Biogenic Amines in Development Parvez, S. & Parvez, H. (eds.). Elsevier/North Holland, Amsterdam.

36. Jones, C. T. & Ashton, I. K. (1976). Biochemical Journal, 154, 149.

37. Salas, M., Vinuela, E. & Sols, A. (1963) Journal of Biological Chemistry, 283, 3535.

38. Faulkner, A. & Jones, C. T. (1976) Archives of Biochemistry and Biophysics, 175, 477.

39. Jones, C. T. & Faulkner, A. unpublished work.

40. Dawes, G. S. & Shelley, H. J. (1968) In Carbohydrate Metabolism and its Disorders. Dickens, W. F., Randle, P. J. & Whelan, W. J. (eds.), Vol. II, pp. 87-121. Academic Press, New York & London.

41. Eisen, H. J., Goldfine, I. D. & Gluisman, W. H. (1973) Proceedings of the National Academy of Sciences, 70, 3454.

42. Jones, C. T. & Ritchie, J. W. K. (1978) Journal of Physiology, 285, 395.

43. Jones, C.T. (1980) In Biogenic Amines in Development, Parvez, S. & Parvez, H. (eds.). Elsevier/North Holland Press, Amsterdam.

44. Beard, A. G. (1975) Journal of Perinatal Medicine, 3, 219.

45. Lagercrantz, H. & Bistoletti, P. (1977) Pediatric Research, 11, 889.

46. Girard, J. R. This symposium.

47. Girard, J. R., Chanez, C., Kervan, A., Tordet-Coridroit, C. & Assan, R. (1976) Biology of the Neonate, 29, 262.

48. Lafeber, H. N., Jones, C. T. & Rolph, T. P. (1979) In Nutrition and Metabolism of the Fetus and Infant. Visser, H. K. A. (ed.), pp 43-62. Martinus Nijhoff Publishers b.v. The Hague.

49. Devos, P. & Hers, H.-G. (1974) Biochemical Journal, 140, 331.

50. Schwartz, A. L. & Rall, T. W. (1973) Biochemical Journal, 134, 985.

51. Bashan, N., Gross, Y., Moses, S. & Gutman, A. (1979) Biochimica et Biophysica Acta, 587, 145.

52. Butcher, F.R. & Potter, V.R. (1972) Cancer Research, 32, 2141.

53. Snell, K. & Walker, D. G. (1973) Biochemical Journal, 134 899.

54. Girard, J. R., Guendet, G. S., Marliss, E.B., Kervan, A., Rieutort, M. & Assan, R. (1973) Journal of Clinical Investigation, 52, 3190.

55. Cornblath, M. & Schwartz, R. (1976) Disorders of Carbohydrate Metabolism in Infancy. W. B. Saunders, Philadelphia.

56. Ballard, F. J. & Oliver, I. T. (1965) Biochemical Journal, 95, 191.

57. Ballard, F. J. & Hanson, R.W. (1967) Biochemical Journal, 104, 866.

58. Schaub, J., Gutman, I. & Lipport, H. (1972) Hormone and Metabolic Research, 4, 110.

59. Girard, J. R.,Cuendet, G. S., Marliss, E. B., Kervan, A., Rieutort, M. & Assan, R. (1973) Journal of Clinical Investigation, 52, 3190.

60. Cowett, R. M., Susa, J. B., Oh, W.(1978) Pediatric Research, 12, 853.

61. Johnston, D. I. & Bloom, S. R. (1973) Archives of Disease in Childhood, 48, 451.

62. Callikan, S., Ferre, P., Pegorier, J.-P. & Girard, J. R. (1979) Journal of Developmental Physiology, 1, 296.

63. Greengard, O. (1977) Pediatric Research, 11, 669.

64. Shelley, H. J. (1961) British Medical Bulletin, 17, 137.

65. Ballard, F. J. and Philippidis, H. (1971) In Regulation of Gluconeogenesis, Soling, H. D. G., Willms, B. (eds.) George Thieme Verlag, Stuttgart.

66. Warnes, D. M., Seamark, R. F. & Ballard, F. J. (1977) Biochemical Journal, 162, 627.

67. Faulkner, A. & Jones, C. T. (1976) Archives of Biochemistry and Biophysics, 176, 171.

68. Jones, C. T. unpublished work.

69. Cornblath, M., Odell, G.B., & Levin, E. Y. (1959), Journal of Pediatrics, 55, 545.

70. Lubchenco, L. O. & Baird, H. (1971) Pediatrics, 47, 831.

71. De Leeuw, R. & De Vries, I. J. (1976) Pediatrics, 58, 18.

72. Haymond, M. W., Karl, I. E. & Pagliara, A. S. (1974) New England Journal of Medicine, 291, 321.

73. Williams, P. R., Fiser, R. H., Sperling, M. A. & Oh, W. (1975) New England Journal of Medicine, 292, 612.

74. Reynolds, J. W. & Mirkin, B. L. (1973) Journal of Clinical Endocrinology and Metabolism, 36, 576.

75. Fiser, R. H. (1976) In Diabetes and other Endocrine Disorders during Pregnancy and in the Newborn. New, M. I. & Fiser, R. H. (eds.), pp 33-50. Alan R. Liss, Inc., New York.

76. Nitzan, M. (1973) Journal of Pediatrics, 83, 505.

77. Ferre, P., Pegorier, J.-P., Marliss, E. B. & Girard, J. R. (1978) American Journal of Physiology, 234, E 129.

78. Ferre, P. This symposium.

79. Soling, H. D., Willms, B., Klineke, J. & Gehlhoff, M. (1970). European Journal of Biochemistry, 16, 289.

80. Soling, H.-D. (1976) In Gluconeogenesis: Its Regulation in Mammalian Species. Hanson, R. W. & Mehlman, M. A. (eds.) pp 369-462. John Wiley & Sons, New York.

81. Dawkins, M. J. R. (1964) Biology of the Neonate, 7, 160.

82. Jones, C. T. & Ritchie, J. W. K. (1978) Journal of Physiology, 285, 395.

83. Bassett, J. M. & Jones, C. T. (1976) In Fetal Physiology and Medicine. Beard, R. W. & Nathanielsz, P. W. (eds.), pp 158-172.

84. Jones, C. T. (1980) In Biogenic Amines in Development. Parvez, S. & Parvez, H. (eds.), Elsevier/North Holland, Amsterdam.

85. Aynsley-Green, A., Lucas, A. & Bloom, S. R. (1979) Acta Paediatrica Scandinavica, 68, 265.

78

86. Lucas, A., Adrian, T. E., Bloom, S. R. & Aynsley-Green, A., (1979) Acata Paediatrica Scandinavica, 68, in press.

87. Kimmel, J. R., Pollock, H. G. & Hayden, L. J. (1978) In Gut Hormones. Bloom, S. R. (ed.), pp 234-241. Churchill Livingstone, Edinburgh.

88. Varma, S., Nickerson, H., Cowan, J. S. & Hetenyi, G. (1973) Metabolism, 22, 1367.

89. Adam, P. A. J., King, K. C. & Schwartz, R. (1968) Pediatrics 41, 91.

90. Cowett, R. M., Susa, J. B., Oh, W. & Schwartz, R. (1978) Pediatric Research, 12, 853.

91. Walker, D. G. & Holland, G. (1965) Biochemical Journal, 97, 845.

92. Faulkner, A. & Jones, C. T. (1976) Archives of Biochemistry and Biophysics, 175, 477.

93. Burch, H. B., Lowry, O. H., Kuhlman, A. M., Skerjance, J., Diamant, E. J., Lowry, S. R. & Von Dippe, P. (1963) Journal of Biological Chemistry, 238, 2267.

94. Dawkins, M. J. R. (1966) British Medical Bulletin, 22, 27.

95. Mersmann, H. J. (1971) American Journal of Physiology, 220, 1927.

96. Bashan, N., Goss, Y., Moses, S. & Grutman, A. (1979) Biochimica et Biophysica Acta, 587, 145.

97. Sperling, M. A., Grajwer, L., Leake, R. D. & Fisher, D. A. (1977) Pediatric Research, 11, 962.

DISCUSSION

<u>Qu.Britton</u> : McCance pointed out that renal function in the
newborn is limited and that this can restrict amino acid
catabolism because of the need to excrete urea and salts.
It is possible that renal function may be inadequate in the
growth retarded foetus and that this may be responsible for
the reduced rate of gluconeogenesis?
What is the blood urea in these animals?

<u>An.Jones</u> : I have no information on renal clearance of urea
in the growth retarded foetus. The ability of the growth
retarded fetal guinea pig to synthesise urea is very much
less than that of normal and its plasma and hepatic ammonia
concentrations are at least 10 times higher than normal.
You are no doubt aware that McCance showed that the fetal
guinea pig at 60 days at least is able to excrete urea and
salts.

<u>Qu.Shelley</u> : Was the low hepatic glucose uptake of the per-
fused fetal guinea pig liver in the presence of other substrates affected by the addition of insulin?
Was the low insulin/glucagon ratio on the growth-retarded
guinea pig due to depressed insulin levels, raised glucagon
or changes in both hormones?

<u>An.Jones</u> : Guinea pig insulin at physiological concentra-
tions causes a small increase in the uptake of glucose in
the presence or absence of other substrates. The low insulin glucagon ratio in the growth retarded fetal guinea pig
was due to a depression of insulin concentrations to about
40 % of their normal value and to raised plasma glucose
concentrations that were very much higher than normal.

<u>Qu.Devos</u> : About the turnover of fetal glycogen (Table 2).
I wonder why there is no synthesis of glycogen during the
period of experimentation. It is than quite normal that
there is no clear incorporation of radioactive glucose and
even it disappears.

<u>An.Jones</u> : There was synthesis of glycogen during the ex-
perimental period although clearly by the nature of the ex-

periment this occurred over a short period. There is no
doubt that the glucose level that is incorporated into glycogen is progressively lost with time, therefore indicating
glycogenolysis.

Qu.Colombo : In the growth retarded guinea pig you showed
elevated values of gluconeogenetic enzymes (PEP-carboxy-
kinase, PEP-carboxylase) but normal alanine-aminotransferase
activity. How do you explain that, particularly taking
also into consideration the secretion in the muscle where
alanine-aminotransferase is increased in the growth retarded animals?
An.Jones : No the data that I showed you, indicated markedly depressed values for the enzymes pyruvate carboxylase
and PEP-carboxylase in the liver of the growth retarded newborn guinea pig. The activity of muscle alanine-aminotransferase was normal.

GLUCOSE TURNOVER
IN THE NEWBORN RAT
Keith Snell

INTRODUCTION

The changes in carbohydrate metabolism that occur during
the course of mammalian development from fetus to adult can
largely be interpreted as an adaptation by the growing
organism to different nutritional environments. The
developmental stages at which major changes in the
nutritional environment occur are at birth, when the
transition is from a transplacental nutrient supply to
maternal milk, and at weaning, when the transition is from
maternal milk to a mixed solid diet. It is at these stages
that major adaptative changes in carbohydrate metabolism
occur and these can be identified by changes in the rate of
flux through a metabolic pathway measured in vivo or, more
indirectly, by the measurements of the maximal capacity of
a pathway in isolated tissue preparations and the activities
of key enzymes in a pathway in vitro. This presentation
will be confined to the fetal-suckling transition and the
adaptive changes in glucose metabolism associated with this
transition. More specifically, we will be concerned with
changes in the rate of glucose utilisation, the rate of
glucose production by gluconeogenesis, and the rate of
glucose production by glycogenolysis. The control of
glucose homeostasis is dependent on the interaction between
these processes.

GLUCOSE METABOLISM IN FETAL AND SUCKLING RATS

It has been suggested that the transplacental supply of
nutrients to the fetus constitutes a high carbohydrate
(specifically glucose) - low fat diet whereas the pattern
of nutrients in the milk supplied to the suckling rat is a
low carbohydrate - high fat diet. The importance of transplacental glucose as an energy fuel for the fetus has been

emphasised and estimated to supply 50 to 80% of the caloric
requirements (1,2). However, more recent work has suggested
that amino acids, which are known to readily cross the
placenta, may make some contribution to oxidative energy,
at least in the sheep (3). This is more apparent during
maternal starvation when the fetal glucose supply is
attenuated (3), but the contribution by amino acids may be
limited in some species, including the rat, by the low
activity of the urea cycle (4). During prolonged maternal
starvation in which ketonaemia is present, there is evidence
that ketone bodies cross the placenta and may be used as
alternative oxidative substrates to glucose (5), although
in the fed mother this is unlikely to be significant. The
case for an energy contribution by free fatty acids is
controversial. Although the placenta is an effective
barrier to the transfer of complex lipids and triglycerides,
some transfer of free fatty acids to the fetus occurs in
certain species (6). It is doubtful whether free fatty
acids serve to any great extent as a source of oxidative
energy since the fetus is engaged for the most part in
synthesising lipid, and measurements of lipogenesis in
vitro and the activity of lipogenic enzymes show an
increased capacity compared to suckling animals (7-9). It
has been demonstrated that the placenta produces lactate
and pyruvate through metabolism of maternal glucose (10,11),
but the extent to which this may contribute to fetal
oxidative energy needs is uncertain. It is clear that the
major contribution to energy requirements in the fetus is
provided by maternal glucose and this is clearly different
from the situation postnatally in the suckling animal.

The contribution by lipid to the caloric content of
milk varies between species, but in all cases predominates
over that supplied by carbohydrate (12). In the suckling
rat, the carbohydrate content of the milk is almost
entirely in the form of lactose and can contribute at the
most 10% of the total required calories (12). Fatty acid
oxidation, and the utilisation of ketone bodies derived
from hepatic fatty acid oxidation, are the major substrates

for oxidative energy metabolism in the suckling rat (13,14).
This is in keeping with the increased circulating concentrations of free fatty acids (Fig. 1) and ketone bodies in
the suckling rat. Unlike the apparently analogous situation
in the starved adult rat, the increased blood concentrations
of fat substrates in the suckling rat are accompanied by
normal blood glucose concentrations relative to the fed
adult value, rather than a decreased level (15). The
enzymes involved in hepatic fatty acid oxidation and ketogenesis and the rates of these processes in vitro are
elevated in the suckling period (14,16).

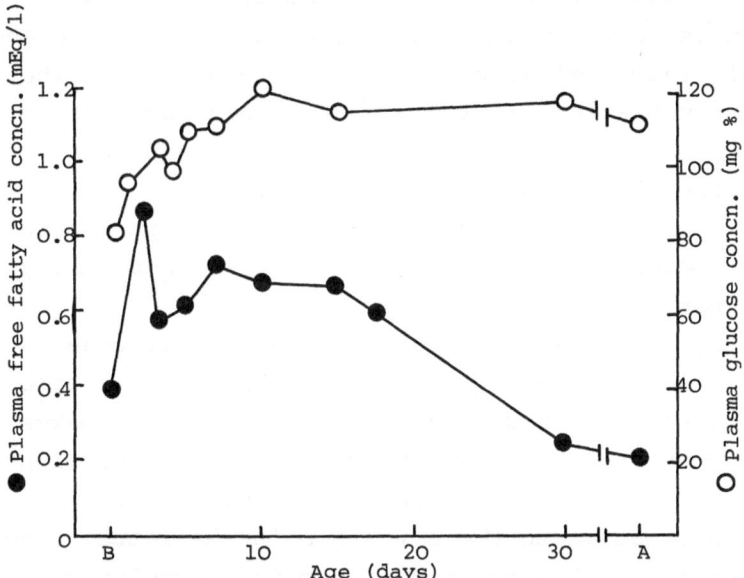

Fig. 1. Plasma free fatty acid and glucose concentrations
during postnatal development. Each point is the
mean of 3-6 observations. A: normal fed adult.

In view of this emphasis on the oxidation of fat-
derived substrates in the suckling rat, the question
arises as to what the significance of glucose metabolism is
in the postnatal animal. The major fates of glucose in the

body are: oxidation by a variety of tissues through glycolysis and the citric acid cycle; lipogenesis by the liver
and adipose tissue; glycogen synthesis by liver and muscle
tissues. The supply of glucose for these purposes derives
from the dietary intake of glucose and the hepatic production
of glucose by glycogenolysis and gluconeogenesis. The
activities of enzymes involved in lipogenesis are decreased
in the liver of suckling rats compared to fetal values,
and only increase again to adult levels when the animals
are weaned from the milk diet on to solid food (16-18).
White adipose tissue depots are low in the suckling rat, but
increase around weaning (19,20). Net liver glycogen
synthesis is very low throughout the suckling period and
again only increases at the time of weaning (21). The
initial accumulation of glycogen at weaning appears to
derive from amino acid carbon (K. Snell, unpublished work),
with glucose playing a permissive role, and is thus
analogous to the fasting-refeeding situation in adults (22).
Muscle glycogen accumulation is not significant during the
suckling period (13).

It thus appears that the major fate of glucose in the
suckling animal is for utilisation in energy production.
However, since the oxidation of lipid substrates makes the
greatest contribution to energy requirements, one might
expect the utilisation of glucose for this purpose to be
diminished relative to the situation in the fetus. Indeed,
the activities of key enzymes of hepatic glycolysis are
decreased from fetal levels during the suckling period and
only increase to adult levels at weaning (17,18). Hepatic
glycolysis, however, is mainly directed towards the
provision of biosynthetic precursors rather than for energy
production per se. The significance of whole body glucose
oxidation can only be assessed from measurements of the
irreversible utilisation of glucose in vivo. For this
purpose, we have employed a combination of $[6-^{14}C]$-glucose
and $[6-^{3}H]$-glucose administered simultaneously to suckling
rats of different ages. Withdrawal of labelled glucose
from the blood glucose pool involves replenishment of the

pool with new unlabelled glucose from the liver to maintain a constant blood glucose concentration. The influx of unlabelled glucose dilutes the radioactivity in the glucose pool and the rate of isotopic dilution therefore provides an indirect measure of glucose utilisation. With $[^{14}C]$-glucose, incomplete glucose oxidation to lactate and alanine in tissues retains isotopically-labelled carbon. The conversion of these precursors to glucose in the liver results in the return of labelled glucose to the pool and therefore an underestimation in the rate of extrahepatic glucose metabolism. By using tritium-labelled glucose, there is extensive exchange of the isotope with protons during glycolysis resulting in a complete loss of label, that is reinforced by additional proton exchange during gluconeogenesis in the liver, so that unlabelled glucose is returned to the pool (Fig. 2). Using $[^{14}C]$-glucose a measure of complete glucose oxidation to CO_2 is provided; with $[^{3}H]$-glucose both oxidative glucose metabolism and partial oxidation by glycolysis is measured. The difference between the rates using the two isotopes is a measure of the recycling of glucose by the Cori and alanine cycles due to glycolysis alone. The choice of the position of tritium label is significant because of the possibility of the reversible exchange of glucose between liver and blood without a net movement of carbon (see Fig. 2). For $[6-^{3}H]$-glucose this will be of little consequence, but for $[2-^{3}H]$-glucose, once phosphorylated, equilibration of glucose 6-phosphate and fructose 6-phosphate results in elimination of tritium from carbon-2. Exchange of $[2-^{3}H]$-glucose across the liver will therefore result in a stripping of isotope from the molecule giving a false impression of apparent net glucose utilisation.

86

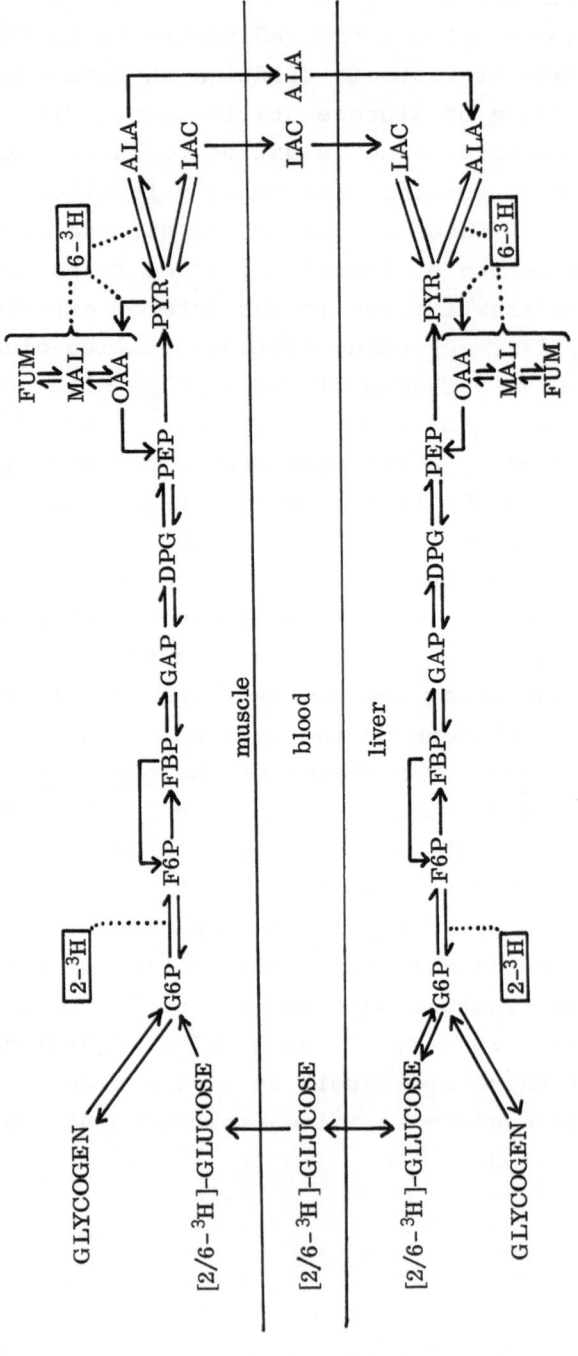

Fig. 2. Loss of radioactivity from differentially-labelled tritiated glucose during glucose utilisation in vivo. Details in the text.

Measurements of glucose utilisation in the suckling rat
show that the rate falls dramatically after birth and is
maintained at a low value until after weaning (Fig. 3).
Recycling of carbon via glycolysis is less than 10% of the
total utilisation at any age, so that the major fate of the
glucose metabolism is complete oxidation (23). The fall
after birth implies that glucose metabolism is greater in
the fetus than the suckling animal and this has been demonstrated for fetal and suckling lambs using $[^{14}C]$-glucose
as a tracer (24).

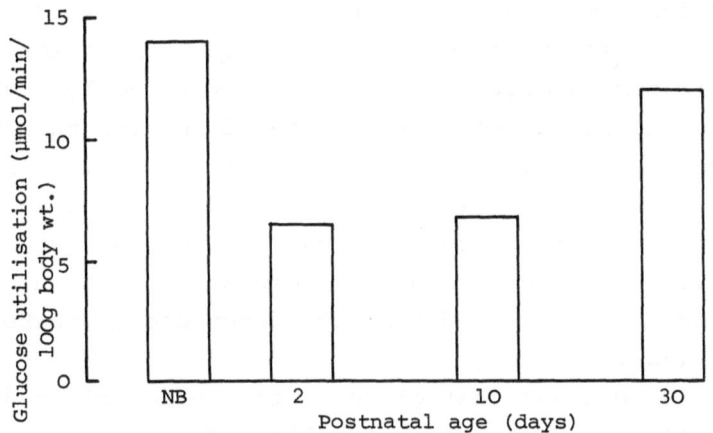

Fig. 3.   Glucose utilisation during postnatal development.
NB, newborn rats. Utilisation was measured in vivo
using $[6-^{3}H]$glucose as a tracer. Data from
Walker & Snell (23).

The reasons for the diminished rate of glucose utilisation in the suckling rat probably include the low
circulating level of insulin (high glucagon: insulin ratio)
and the utilisation of fatty acids by muscle tissues which,
at least in the adult, is known to inhibit glucose utilisation (25,26). In addition, the major tissue involved in
glucose oxidation in the adult is the brain and in the
suckling rat the adaptation by the brain to the use of
ketone bodies as an alternative oxidisable substrate (27,28)
is a major factor in the diminished body glucose require-

ments.

Despite the use of fat-derived fuels for oxidative metabolism by a number of tissues in the suckling animal, certain tissues have an obligatory requirement for glucose for energy metabolism, e.g. the blood cells, renal medulla, intestinal mucosa, etc. (29). In addition, it has been shown that, in the newborn dog at least, heart muscle is dependent largely on glycolysis rather than fatty acid oxidation for energy (30). Even the brain which partially adapts to the utilisation of keton bodies as an energy substrate, apparently only derives about 30% of its oxidative metabolism from this source in the 18-day-old rat (27). At earlier ages the quantitative utilisation of ketone bodies is very uncertain, but is likely to be less than the value in the 18-day-old rat because of the immaturity of the necessary enzymic apparatus (28,31). Studies in man using non-radioactive isotopically-labelled glucose have, in fact, shown direct correlations between the rates of body glucose production (and therefore utilisation) and estimated brain weight during childhood and in newborn infants (32).

It is apparent that glucose utilisation plays a significant role in the energy metabolism of the suckling neonate, a finding that one might suspect in view of the fact that blood glucose concentrations in the suckling rat are not reduced as in the starved adult animal. However, calculations of the maximal contribution that dietary carbohydrate present in milk can make to glucose formation show that this is insufficient to balance the measured rates of glucose utilisation (33). This requires a significant conversion of non-carbohydrate precursors to glucose by hepatic gluconeogenesis. Net breakdown of glycogen does not occur after the immediately postnatal period and is unable to contribute to the total body glucose pool (23,33). In fact measurements of gluconeogenesis from lactate in vivo and in the perfused liver preparation in vitro show that this metabolic process is enhanced in the suckling rat (Fig. 4). Similar conclusions were reached by Bier et al. (32) from their in vivo studies on human infants.

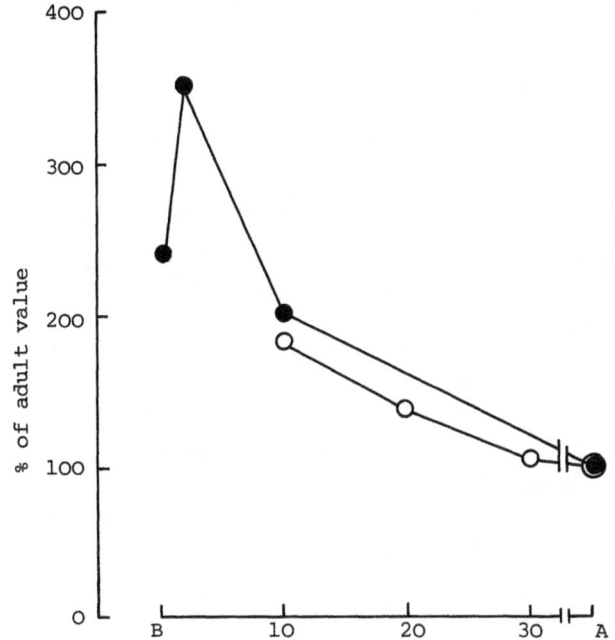

Fig. 4. Gluconeogenesis from lactate during postnatal
development. Gluconeogenesis was measured in
vivo (●) from ^{14}C lactate and in the perfused
liver in vitro (O) from 10 mM-lactate. Data
taken, in part, from Walker & Snell (23) and
Snell (34).

The importance of hepatic gluconeogenesis in maintaining blood glucose concentrations is shown by an experiment in which a potent inhibitor of gluconeogenesis,
3-mercaptopicolinate, was injected into 10-day-old rats
in vivo and produced a marked fall in glucose concentration
and an accumulation of the gluconeogenic precursor lactate
in the blood (Fig. 5).

The maintenance of an increased rate of hepatic
gluconeogenesis in the suckling rat is due to a number of
factors. One of these is the continued supply of fatty
acids in the diet (60), because their
oxidation provides not only reduced nucleotide cofactors
(NADH) which are required in the gluconeogenic pathway,
but also acetyl CoA which is an obligatory activator of one

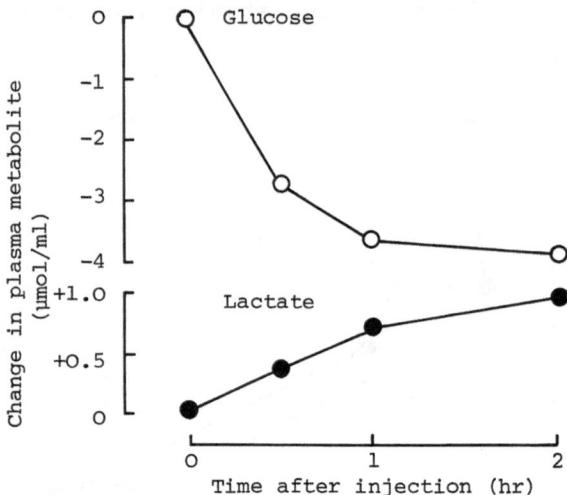

Fig. 5. Effect of 3-mercaptopicolinate on plasma
glucose and lactate in suckling 10-day-old
rats. Mercaptopicolinate was injected i.p.
at a dose of 10 mg/100g body wt. in 0.9%
saline. Each point is the mean of 4 observations.

of the key enzymes in the pathway (pyruvate carboxylase).
Another factor is the hormonal status of the animal and
the high plasma glucagon, low insulin levels in the suckling rat (36-38) are appropriate for this purpose since
they result in a high concentration of hepatic cyclic AMP
(37-39) which stimulates gluconeogenesis. A further factor
involved in maintaining gluconeogenesis in the suckling rat
is the elevated activities of the key enzymes of gluconeogenesis compared to the fetus and adult (17), and the
increased level of mitochondrial pyruvate carboxylation
(40), which determine the maximal flux through the pathway.

GLUCOSE METABOLISM IN THE NEWBORN RAT
At birth the transition from the fetal pattern of carbohydrate
metabolism to that characteristic of the suckling rat must
occur. In the present context we are concerned with the
timing of the adaptations in glucose utilisation, gluconeogenesis and glycogenolysis, and the factors that control
these adaptations.

Table 1 shows the main factors involved in regulating gluconeogenesis.

Table 1.   *Factors Regulating Gluconeogenesis*

1. Presence of gluconeogenic enzymes.

2. Availability of precursors.

3. Adequacy of energy supply (e.g. fatty acid oxidation; citric acid cycle functional; oxidative phosphorylation proceeding).

4. Generation of reducing power (e.g. fatty acid oxidation)

5. Mitochondrial anion translocation (e.g. pyruvate transport; aspartate and malate translocation; ATP translocation).

6. Hormonal status (e.g. glucagon ↑, insulin ↓; glucocorticoids ↑).

In the rat at birth the levels of all the gluconeo-genic enzymes, except PEP carboxykinase, are present at about the adult value (17,41). PEP carboxykinase is very low at birth but rapidly increases in activity and this is essential for the maximal operation of the patway (42). In other species, such as the guinea pig and man, PEP carboxy-kinase as well as the other gluconeogenic enzymes are present already in substantial amounts at birth. But even for these species there is a rapid further increase in PEP carboxykinase which appears to be important for the gluco-neogenic pathway to operate at its maximum rate (43-45). The second factor that regulates the rate of gluconeogenesis at birth is the availability of endogenous gluconeogenic precursors, of which lactate, alanine and perhaps glycerol are the most significant. Thirdly, an adequate energy supply is required to support gluconeogenesis, and this involves substrate oxidation (particularly fatty acids for the suckling rat), the operation of the citric acid cycle and the functioning of oxidative phosphorylation. Indeed, Ballard (46) has shown that rats at the moment of birth have low hepatic ATP/ADP ratios which increase during the

first 30 minutes after birth once breathing is properly
established and tissues are fully oxygenated. The
establishment of normal ATP/ADP ratios is also probably
related to a mitochondrial maturation of adenine nucleotide
transport that occurs, apparently as a result of tissue
oxygenation, after birth (47). It is these factors which
probably limit the operation of gluconeogenesis in utero,
even in those species which already possess the necessary
enzymes [see e.g. Warnes et al. for data on lambs (48)].

The fourth regulating factor for gluconeogenesis is
the provision of reducing power required in the pathway and
generated by fatty acid oxidation in the main, although
utilisation of lactate for glucose formation will produce
its own NADH in stoichiometric amounts through the lactate
dehydrogenase reaction. Fifthly, mitochondrial anion translocation is essential for gluconeogenesis and differential
permeability properties are acquired soon after birth (47,
49). Because of the mitochondrial localisation of the
gluconeogenic enzyme, pyruvate carboxylase, transport of
pyruvate into the mitochondria is required and this is
increased substantially during the first hour after birth in
the rat (40). Finally, gluconeogenesis is stimulated by an
appropriate hormonal status - raised glucagon levels,
lowered insulin levels and a raised glucocorticoid level -
which exerts its effects at one or more of the regulatory
loci already referred to above.

We will now consider some of these regulatory factors
in the newborn rat with particular attention to the timing
of the responses. Plasma glucose levels which are high at
birth, perhaps due to neural activation of limited glycogenolysis at the moment of placental detachment, fall to a
low level by 90 minutes and then there is a recovery to
normoglycaemia by about 3 hours post partum (Fig. 6). [All
animals were delivered by caesarian section (after maternal
cervical dislocation) on the final day of gestation to
synchronise birth for all members of the litter and allow
the accurate timing of postnatal events.] The transient
hypoglycaemia, which is a common finding in most species,

Fig. 6. Plasma glucose concentration after surgical
delivery of term fetal rats. Each point is
the mean of 5-8 observations. Bars show S.E.M.

including man, is the result of a high rate of body glucose
utilisation at birth, as in the fetus, and is also perhaps
associated with the increased extrauterine muscular activity.
Glucose utilisation must obviously exceed endogenous production at this time, and the reverse must occur during the
recovery from hypoglycaemia. The sources of endogenous
glucose production may be gluconeogenesis and/or glycogenolysis.

Fig. 7 shows that plasma lactate is high at birth, as
in the fetus, and may additionally result from anaerobic
muscular activity. This particularly applies to heart
muscle which seems to be dependent at birth on glycolysis
for its energy requirements (51-53). Lactate is an important precursor for gluconeogenesis and the blood level
falls sharply from 30 minutes after birth to reach a low
plateau by about 2 hours. Similarly the level of another
gluconeogenic precursor, alanine, falls after birth to
reach a low plateau level by 2 hours. It seems that endogenous precursors for gluconeogenesis are used soon after
birth but that their availability is diminished after a few
hours.

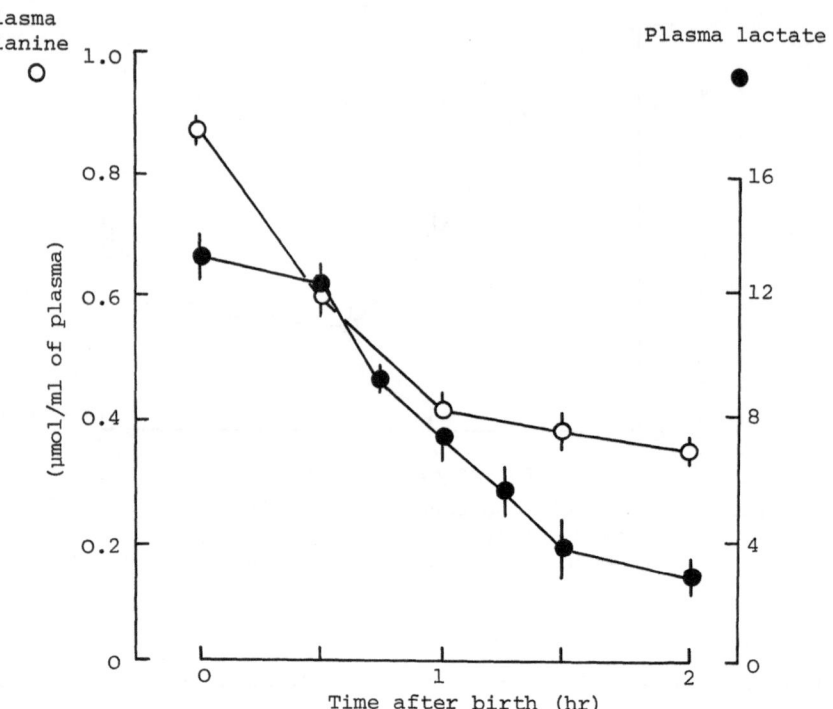

Fig. 7.   Plasma concentrations of lactate and alanine
after surgical delivery of term fetal rats.
Each point is the mean of 3-6 observations.
Bars show S.E.M.

Plasma glucagon levels rise sharply after birth to
peak at 30 minutes and this is appropriate for a stimulation
of gluconeogenesis (54).   The initiation of gluconeogenesis
at this point (30 minutes) is also suggested by the fall in
blood gluconeogenic precursors at this time and the
increased liver ATP/ADP ratios and mitochondrial maturation
referred to above.   Plasma insulin levels, which are high
at birth and are therefore appropriate for the high rate of
glucose utilisation, fall after birth to reach a low plateau
level between 1 and 2 hours (54).   The rise in glucagon at
30 minutes is probably the most significant factor in stimulating gluconeogenesis after birth.   Glucagon is known to
stimulate the transport of pyruvate into mitochondria (55)
(which occurs within the first hour after birth);   it is
known to activate a component of the electron transport

chain and so stimulates the phosphorylation of adenine nucleotide (56);
and, in the rat at least, it induces the appearance of PEP carboxykinase - the rate-limiting enzyme for gluconeogenesis (57). The injection of glucagon into newborn rats at the time of delivery stimulates
the rate of glucose production, the rate of lactate utilisation, and
the incorporation of $[^{14}C]$-lactate into blood glucose in vivo (58). The
stimulation of gluconeogenesis by these pharmacological doses of glucagon abolished the postnatal phase of hypoglycaemia (58). Thus, the
natural, glucagon-mediated, initiation of gluconeogenesis in the first
hour is one of the important factors in establishing glucose homeostatic
controls after birth.

A way of assessing gluconeogenesis in vivo is to inject an animal
with a radioactive dose of labelled glucose in quantities small enough
not to disturb the actual size of the glucose pool. The injected
labelled glucose distributes itself through the glucose pool and its
dilution by inflowing unlabelled glucose, newly synthesised from gluconeogenic precursors, provides a measure of the rate of gluconeogenesis,
a technique already referred to for measuring glucose utilisation. The
two most commonly used glucose tracers are carbon-2 tritiated glucose
and carbon-6 tritiated glucose and the tritium at these two positions
is lost at different stages of glycolysis or gluconeogenesis as indicated above (Fig.2). The most significant difference that this makes
is that exchange of $[2-^{3}H]$-glucose between the blood and liver results
in loss of label, so that when this glucose returns to the blood pool
it appears as unlabelled glucose. This introduces a recycling error
which causes an overestimation of glucose production with this isotope
(Fig. 8).

Another difference between the two labelled glucose molecules is
also apparent when we consider exchange of glucose between the blood
and the liver. For both types of glucose, lactate and alanine precursors from muscle will be unlabelled and will give unlabelled glucose
to dilute the blood pool (the measure of gluconeogenesis). For carbon-
2 tritiated glucose, additional unlabelled glucose will be formed
through recycling. In addition, in the presence of glycogen turnover,
carbon-6 tritiated glucose entering the liver will pass into the glucose
6-phosphate pool and will appear in glycogen in a labelled form. Subsequent glucose production by net glycogenolysis will produce predominantly labelled glucose molecules so that glycogen breakdown would

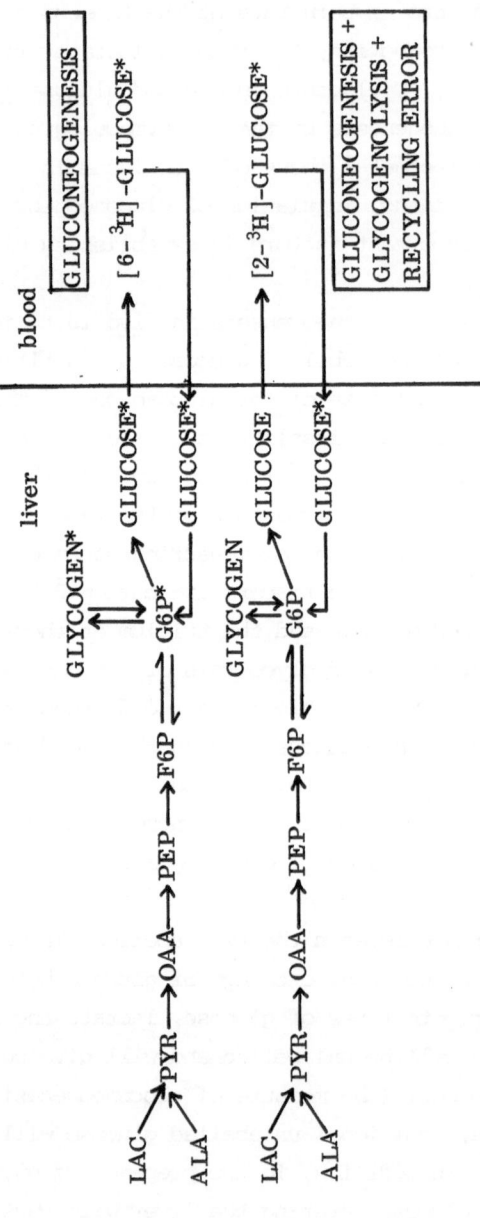

Fig. 8.   Use of differentially-labelled glucose molecules in determining glucose
production in vivo.   Further details are in the text.

not be detected by isotopic dilution of the blood glucose
pool.   Of course massive glycogen breakdown would be
detected because the labelled glycosyl units will tend to
be confined to the outer newly synthesised branches of the
glycogen molecule .   For carbon-2 tritiated glucose, label
is lost in the hexose monophosphate pool and the glycogen
formed will be unlabelled.   If glycogen breakdown occurs,
unlabelled glucose molecules will be produced which will
dilute the labelled blood glucose pool.   Thus, with $[2-{}^{3}H]$ -
glucose, total glucose production from both gluconeogenesis
and glycogenolysis will be measured, but there will be an
overestimation due to the recycling error.   With $[6-{}^{3}H]$ -
glucose, gluconeogenesis alone will be measured.

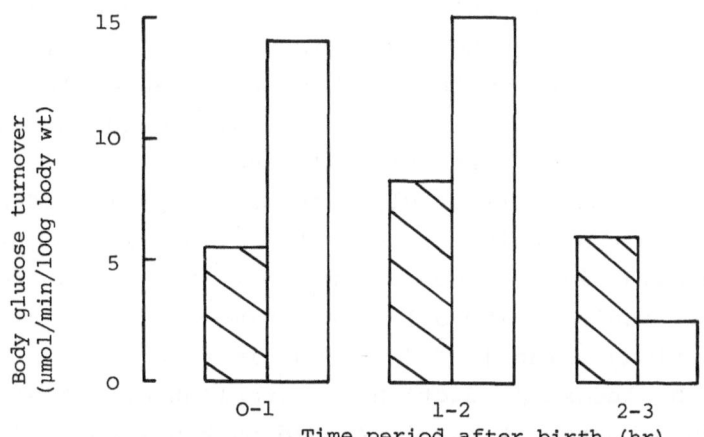

Fig. 9.   Glucose turnover rates at intervals after birth.
Open columns represent rates of glucose utilisation
and hatched columns represent rates of glucose
production.   Data taken from Snell & Walker (50).

By combining equations for the decline in $[6-{}^{3}H]$ -
glucose specific radioactivity with those for changes in
plasma glucose concentration itself, it is possible to
calculate both the rate of glucose production and the rate
of glucose utilisation (50).   These rates were determined at
intervals of 0 to 1 hour after birth, 1 to 2 hours, and 2 to
3 hours (Fig. 9).   Gluconeogenesis is initiated in the first
hour, but the rate of glucose utilisation exceeds this value

considerably and this accounts for the rapid fall in blood
glucose over this period. A high rate of glucose utilisation at birth was also suggested by De Meyer (59) on other
indirect evidence, and is favoured by the high plasma insulin
concentration at this time (54). Indeed, as the insulin
falls to reach a low level by 2 hours (54), then the rate of
glucose utilisation falls (Fig. 9). Another factor
favouring a decrease in glucose utilisation in the first few
hours after birth is the rise in plasma concentrations of
free fatty acids (50,60), and their oxidation to ketone
bodies which occurs at about 2 hours post partum (50,60),
both of which provide alternative oxidative energy fuels.
Eventually utilisation falls below the rate of gluconeogenesis which marks the upturn in blood glucose concentration. During this initial period after birth, the rate
of gluconeogenesis is falling off as gluconeogenic precursors become diminished. What is clear is that in the
first hours after birth there is no appreciable glycogen
breakdown, because massive glycogenolysis would have been
revealed even using 6-tritiated glucose.

Measurements of liver glycogen concentration after
delivery show that this is constant for the first hour,
decreases slightly by 2 hours, and only then begins to
decrease appreciably (50,54). This time-course is supported
by observations using $[2-^3H]$-glucose in vivo (61). Soon
after birth glucose production is established, but at 4
hours (and 6 hours) it has increased markedly and this is
at a time when glycogen concentrations are now falling
rapidly. It is this breakdown of glycogen and the accom-
panying increase in glucose production which is one of the
main factors in rapidly restoring normal glucose concentrations after birth and maintaining these levels until
gluconeogenesis from exogenous dietary sources is
established. By the time glycogen stores are exhausted
(about 12 hours after birth dietary-supported gluconeogenesis is actively proceeding. The utilisation of dietary-
supplied fat substrates for body oxidative metabolism is
also established by this time (60), and will have partially

replaced the use of glucose as an energy substrate.

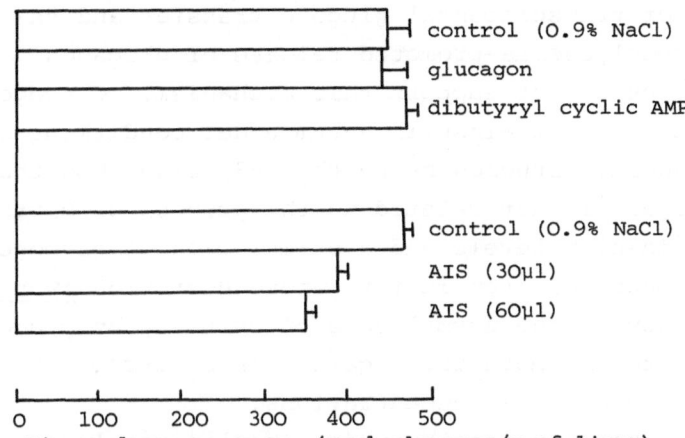

Liver glycogen concn. (μmol glycogen/g of liver)

Fig. 10.   Liver glycogen breakdown in newborn rats.
Surgically-delivered rats were injected with
glucagon (50μg/rat), dibutyryl cyclic AMP (0.25
mg/rat), or guinea-pig anti-insulin serum (AIS)
at 30 μl or 60 μl/rat. Animals were killed by
decapitation one hour later. Data from Snell &
Walker (58,62).

The question arises of why, if plasma glucagon peaks at
30 minutes, glycogen breakdown does not occur until after 2
hours. Exogenous glucagon injected at birth has no apparent
effect on liver glycogen levels one hour later and this
suggests that the liver is unresponsive to glucagon at this
time (Fig. 10). It is worth noting that the presumed second
messenger for glucagon action (cAMP) is also ineffective.
One reason for this may be that insulin levels are high
immediately after birth and do not fall to low levels until
1 to 2 hours (54), and may be inhibiting glycogen breakdown
due to effects on glycogen synthetase and phosphorylase.
When rats were injected at birth with anti-insulin serum
glycogen breakdown was readily stimulated within one hour
in a dose-related fashion (Fig. 10). The breakdown of
glycogen induced by anti-insulin serum abolished the hypoglycaemia seen in untreated animals at 1 hour after birth
(62). It seems that the fall in plasma insulin after birth
is the most significant factor in initiating postnatal
glycogenolysis.

It has been stated that the breakdown of liver glycogen in the newborn rat occurs as a result of birth, the cessation of transplacental glucose transfer and the subsequent hypoglycaemia-promoted release of glucagon. The above studies do not support this mechanism. Glycogenolysis was promoted by anti-insulin serum under conditions of hyperglycaemia. Studies by Portha (63) also show that glycogenolysis is not related to the process of birth, since a fall in insulin levels and a stimulation of glycogen breakdown occurred even in rats whose gestation in utero was prolonged beyond the normal date of delivery by progesterone injected into the mothers. Again this occurred without any fall in blood glucose concentrations.

As to the mechanism by which insulin regulates glycogen metabolism at birth, clearly a fall in insulin will favour the inactivation of glycogen synthetase (by promoting interconversion of a and b forms). So there would be a decrease in glycogen concentration in the face of continued turnover. In addition, a fall in insulin may promote the activation of glycogen phosphorylase to stimulate glycogen breakdown. This action of insulin is not related to cyclic AMP, since this agent is without effect on glycogenolysis in the first hour after birth (58). This suggests an action of insulin that by-passes the cyclic AMP-protein kinase control of the glycogenolytic cascade, and which perhaps involves antagonism of the α-adrenergic-mediated activation of phosphorylase that is believed to operate through a rise in cytoplasmic calcium ion concentration which stimulates phosphorylase kinase (64,65).

Fig. 11 summarises the timing of the major processes involved in controlling glucose homeostasis in newborn rats. At birth there may be a limited breakdown of glycogen as a result of sympathetic nervous activity associated with the stress of parturition. However, the high rate of glucose utilisation, in the absence of significant glucose production, results in a sharp drop in blood glucose levels. At 30 minutes a peak of plasma glucagon initiates gluconeogenesis and over the next hour blood gluconeogenic pre-

101

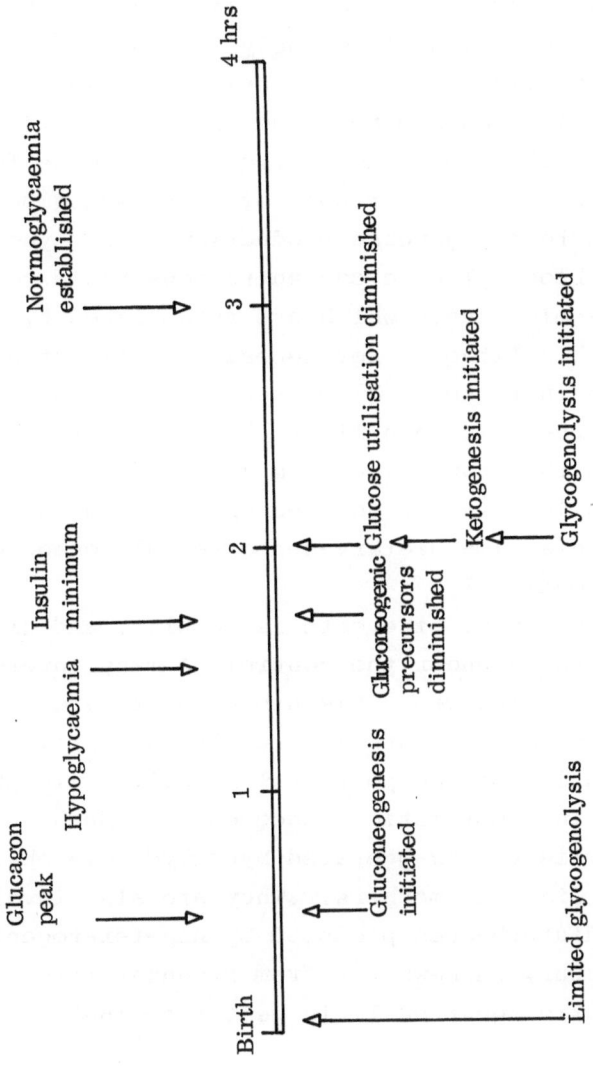

Fig. 11. Timing of postnatal events involved in the control of glucose homeostasis in the newborn rat

cursors are rapidly utilised, slowing down the fall in blood glucose so that it plateaus at a minimal level at about 90 minutes after birth. During this time plasma insulin levels are falling from the high values at birth and reach a minimum at about the time that the availability of gluconeogenic precursors is diminished. This low insulin concentration initiates the phase of rapid glycogenolysis, and also decreases glucose utilisation. Fatty acid oxidation is also activated at this time (as shown by the initiation of ketogenesis) and this will also diminish the requirements for body glucose utilisation. The result of decreased glucose utilisation and increased glucose production from glycogenolysis is that blood glucose concentrations begin to rise towards normoglycaemic levels which are established by about 3 hours after birth. Although ketogenesis is initiated at 2 hours, the major increase in blood ketone bodies occurs after 12 hours at the time when liver glycogen stores have been exhausted and the transition from predominantly glucose oxidation to predominantly fat-derived substrate oxidation is complete. By this time dietary-supported gluconeogenesis is also operating maximally.

Clearly this sequence of events is complex, and glucose homeostasis at birth is under the control of many interacting factors and influences. Disturbances or defects in any of these processes or events may result in a prolongation or exacerbation of the postnatal hypoglycaemic phase that could result in symptomatic consequences. Such disturbances underlie the exaggerated hypoglycaemia observed in the infants of diabetic mothers. They are also involved in the metabolic disturbances produced by non-teratogenic doses of methylmercury in newborns from pregnant rats exposed to this toxic agent early in gestation (66).

REFERENCES

1. Adam, P. A. J. (1971) Advan. Metab. Disord. 5, 183-275.
2. Battaglia, F. C. & Meschia, G. (1978) Physiol. Rev. 58, 499-527.
3. Battaglia, F. C. (1979) Ciba Found. Symp. 63, 57-68.
4. Raiha, N. C. R. (1971) in The Biochemistry of Development (Benson, P. F. ed.), pp. 141-160, William Heinemann, London.

5. Shambaugh, G. E., Mrozak, S. C. & Freinkel, N. (1977) Metab. Clin. Exp. $\underline{26}$, 623-636.

6. Hull, D. & Elphick, M. C. (1979) Ciba Found. Symp. $\underline{63}$, 75-86.

7. Taylor, C. B., Bailey, E. & Bartley, W. (1967) Biochem. J. $\underline{105}$, 717-722.

8. Jones, C. T. (1976) Biochem. J. $\underline{156}$, 357-365.

9. Jones, C. T. & Ashton, I. K. (1976) Biochem. J. $\underline{154}$, 149-158.

10. Freinkel, N. (1965) in On the Nature and Treatment of Diabetes (Leibel, B. S. & Wrenshall, G. A., eds.), pp. 679-691, Excerpta Medica, Amsterdam.

11. Burd, L. I., Jones, M. D., Simmons, M. A., Makowski, E. L., Meschia, G. & Battaglia, F. C. (1975) Nature (Lond.) $\underline{254}$, 710-711.

12. Walker, D. G. (1971) in The Biochemistry of Development (Benson, P.F. ed.), pp. 77-95, William Heinemann, London.

13. Hahn, P. & Koldovsky, O. (1966) The Utilisation of Nutrients during Postnatal Development, Pergamon Press, Oxford.

14. Bailey, E. & Lockwood, E. A. (1973) Enzyme $\underline{15}$, 239-253.

15. Yeh, Y. Y. & Zee, P. (1976) Pediat. Res. $\underline{10}$, 192-197.

16. Williamson, D. H. & Buckley, B. M. (1973) in Inborn Errors of Metabolism (Hommes, F. A. & Van den Berg, C. J., eds.), pp.81-92, Academic Press, London.

17. Vernon, R. G. & Walker, D. G. (1968) Biochem. J. $\underline{106}$, 321-330.

18. Vernon, R. G. & Walker, D. G. (1968) Biochem. J. $\underline{106}$, 331-338.

19. Cryer, A. & Jones, H. M. (1978) Biochem. J. $\underline{172}$, 319-325.

20. Cryer, A. & Jones, H. M. (1979) Biochem. J. $\underline{172}$, 711-724.

21. Ballard, F. J. & Oliver, I. T. (1963) Biochim. Biophys. Acta $\underline{71}$, 578-588.

22. Hems, D. A., Whitton, P. A. & Taylor, E. A. (1972) Biochem. J. $\underline{129}$, 529-538.

23. Walker, D. G. & Snell, K. (1973) in Inborn Errors of Metabolism (Hommes, F. A. & Van den Berg, C. J., eds.), pp. 97-116, Academic Press, London.

24. Warnes, D. M., Seamark, R. F. & Ballard, F. J. (1977) Biochem. J. $\underline{162}$, 617-626.

25. Randle, P. J., Garland, P. B., Hales, C. N. & Newsholme, E. A. (1966) Recent Progr. Horm. Res. $\underline{22}$, 1-48.

26. Ruderman, N. B., Goodman, M. N., Berger, M. & Hagg, S. (1977) Fed. Proc. $\underline{36}$, 171-176.

27. Cremer, J. E., Teal, H. M. & Heath, D. F. (1975) in Normal and
    Pathological Development of Energy Metabolism (Hommes, F. A. &
    Van den Berg, C. J., eds.), pp. 133-141, Academic Press, London.

28. Land, J. M. & Clark, J. B. (1975) in Normal and Pathological
    Development of Energy Metabolism (Hommes, F. A. & Van den Berg,
    C. J., eds.), pp. 155-165, Academic Press, London.

29. Krebs, H. A. (1972) Advan. Enz. Regul. 10, 397-420.

30. Clark, C. M. (1973) Diabetes 22, 41-49.

31. Page, M. A., Krebs, H. A. & Williamson, D. H. (1971) Biochem. J.
    121, 49-53.

32. Bier, D. M. et al. (1977) Diabetes 26, 1016-1023.

33. Snell, K. & Walker, D. G. (1973) Enzyme 15, 40-81.

34. Snell, K. (1975) in Normal and Pathological Development of Energy
    Metabolism (Hommes, F. A. & Van den Berg, C. J., eds.), pp.77-94,
    Academic Press, London.

35. Ferre, P., Pegorier, J. P. & Girard, J. (1978) in Diabetes, Obesity
    and Hyperlipidemias (Crepaldi, G., Lefebvre, P. J. & Alberti, K.G.
    M.M., eds.), pp. 11-19, Academic Press, London.

36. Girard, J. R., Kervran, A. & Assan, R. (1975) in Early Diabetes in
    Early Life (Camerimi-Davalos, R. A. & Cole, H. S., eds.), pp.57-71,
    Academic Press, New York.

37. Blazquez, E., Sugase, T., Blazquez, M. & Foa, P. P. (1974) J.Lab.
    Clin. Med. 83, 957-967.

38. Beaudry, M. A., Chiasson, J. L. & Exton, J. H. (1977) Am. J. Physiol.
    233, E175-E180.

39. Stratford, M. R. L. & Snell, K. (1979) Biochem. Soc. Trans. 7,
    902-904.

40. Snell, K. (1974) Int. J. Biochem. 5, 463-469.

41. Ballard, F. J. & Hanson, R. W. (1967) Biochem. J. 104, 866- 871.

42. Hanson, R. W., Reshef, L. & Ballard, F. J. (1975) Fed. Proc. 34,
    166-171.

43. Jones, C. T. & Ashton, I. K. (1976) Arch. Biochem. Biophys. 174,
    506-522.

44. Robinson, B. H. (1976) Biol. Neonate 29, 48-55.

45. Marsac, C., Saudubray, J. M., Moncion, A. & Leroux, J. P. (1976)
    Biol. Neonate 28, 317-325.

46. Ballard, F. J. (1971) Biochem. J. 124, 265-274.

47. Sutton, R. & Pollak, J. K. (1978) Differentiation 12, 15-21.

48. Warnes, D. M., Seamark, R. F. & Ballard, F. J. (1977) Biochem. J. 162, 627-634.

49. Pollak, J. K. (1975) Biochem. J. 150, 477-488.

50. Snell, K. & Walker, D. G. (1973) Biochem. J. 132, 739-752.

51. Hoerter, J. A. (1976) Pfluegers Arch. 363, 1-6.

52. Hoerter, J. A. & Opie, L. H. (1978) Biol. Neonate 33, 144-161.

53. Jarmakani, J. M., Nakazawa, M., Nagatomo, T. & Langer, G. (1978) Am. J. Physiol. 235, H469-H474.

54. Girard, J. R., Cuendet, G. S., Marliss, E. B., Kervran, A., Rieutort, M. & Assan, R. (1973) J. Clin. Invest. 52, 3190-3200.

55. Titheradge, M. A. & Coore, H. G. (1976) FEBS Lett. 63, 45-50.

56. Titheradge, M. A., Binder, S. B., Yamazaki, R. K. & Haynes, R. C. (1978) J. Biol. Chem. 253, 3357-3360.

57. Philippidis, H. & Ballard, F. J. (1970) Biochem. J. 120, 385-392.

58. Snell, K. & Walker, D. G. (1973) 134, 899-906.

59. De Meyer, R., Gerard, P. & Verellen, G. (1971) in Metabolic Processes in the Foetus and Newborn Infant (Jonxis, J. H. P., Visser, H. K. A. & Troelstra, J. A., eds.), pp. 281-291, Stenfort Kroese N. V., Leiden.

60. Ferre, P., Pegorier, J. P., Williamson, D. H. & Girard, J. R. (1978) Biochem. J. 176, 759-765.

61. Girard, J. R. & Guillet, I. (1975) Biochem. J. 148, 345-347.

62. Snell, K. & Walker, D. G. (1978) Diabetologia 14, 59-64.

63. Portha, B., Rosselin, G. & Picon, L. (1976) Diabetologia 12, 429-436.

64. Van de Werve, G., Hue, L. & Hers, H.-G. (1977) Biochem. J. 162, 135-142.

65. Van de Werve, G., Hue, L. & Hers, H.-G. (1977) Biochem. J. 162, 143-146.

66. Snell, K., Ashby, S. L. & Barton, S. J. (1977) Toxicology 8, 277-283.

DISCUSSION

Qu.Ferre : How can you reconciliate the facts 1)that you find a very low glucose recycling and 2) that you think that lactate is an important glucose preaursor?

An.Snell: Immediately after birth there is no question that lactate is significant endogenous precursor for glucose synthesis. However the question arises as to the significance of lactate for gluconeogenesis once suckling is established. We can discount dietary lactate in the milk as a significant source and so lactate must arise endogenously. There is no net glycogenolysis during the suckling period (after the first few days post partum) and the lactate must be produced from circulating glucose and its conversion to glucose therefore constitutes a recycling of carbon. As I have shown (Fig.4), the conversion of lactate to glucose, estimated either from radioactive incorporation in vivo or net synthesis in vitro, is considerably elevated in the suckling rat compared to the adult. In addition, inhibition of gluconeogenesis in vivo results in an accumulation of blood lactate and a fall in blood glucose (Fig.5). Altogether this data suggests that lactate recycling to glucose is an important mechanism for maintaining glucose homeostasis during suckling. As I pointed out, by using a combination of $[6-^3H]$ - glucose and $[^{14}C]$ - glucose,it should be possible to estimate the extent of recycling of glucose carbon through lactate (and alanine) (Fig.2). With this method, it appears that recycling is negligible in the two-day-old suckling rat (Snell,K., and Walker,D.G., (1973) Biochem.J.132,739-752) or at later ages during suckling (Vernon,R.G. and Walker, D.G., (1972) Biochem.J. 127,521-529). In addition, using these estimates of recycled glucose, it appears that there is a large discrepancy between net glucose utilisation and the calculated contribution of exogenous precursors to glucose formation (Snell,K. and Walker,D.G. (1973) Enzyme 15,40-81; Snell,K., (1980) in "Biochemical Development of the Fetus and the Neonate" (Jones,C.T. ed.), Vol.2,Elsevier, Amsterdam, in press). These various discrepancies suggest that the question of recycling should be reassessed and that perhaps there is an error in using the $[^3H]$ - glucose - $[^{14}C]$ - glucose difference method for this purpose. One possibility is that $[^3H]$ elimination from recycled precursors (such as lactate) might be incomplete. Oxidation of $[^3H]$ - glucose could result in formation of ^3H-labelled NADH which may be reincorporated back into glucose at the glyceraldehyde 3-phosphate dehydrogenase step. Although this might be considered unlikely, intramitochondrial ^3H-labelled NADH has been shown to promote tritium incorporation into glucose (Rognstad, R. and Katz,J., (1971) FEBS Lett. 15,219-222). Measurements of $[^{14}C]$ - glucose turnover and $^{14}CO_2$ production in vivo may help to resolve this problem.

Qu.Ferre : It may be important to precise at which temperature the studies are conducted as the newborn rat is not homeotherm (30°C is already a cold exposure).
An.Snell: I agree that it is essential in neonatal studies to state the temperature at which the animals are kept. In all our studies where the neonates have been removed from their mothers, they are kept in a humid atmosphere at a temperature such that the ambient temperature for the rats is 35°C. This is the temperature that we have observed for the rat nest when the mother is suckling her pups.

Qu.Britton : Your values for blood lactate in the newborn rat delivered by caesarian section seem very high and must indicate a very substantial acidosis. Are the lactate values as high in vaginal delivery?
An.Snell: I agree that the blood lactate values are indeed high in the newborn rat, as they are in the fetus. This is in part due to the high rate of fetal glycolysis, which continues in the early newborn period, but also to the vigorous muscular activity associated with extrauterine life which will result in muscle glycogen breakdown and lactate formation. Prior to the initiation of gluconeogenesis and metabolism of lactate by the liver, there will be an accumulation of lactate in the blood. Essentially

108

similar blood lactate concentrations are found at birth
regardless of whether rats are delivered by caesarian section (following either cervical dislocation or light ether
anaesthesia of the dam) or in rats spontaneously born by
vaginal delivery. The same is true for intracellular lactate concentrations in newborn rat liver (Ballard, F.J.,
(1971) Biochem.J. 124, 265-274).

Qu.Minkowski : Reconversion of lactate into glucose in the
liver in hypoxic state.
An.Snell: Blood lactate concentrations, which are high at
birth, show a 30 minute lag after delivery before they
begin to fall (Fig.7). This is interpreted as a delay in
hepatic utilisation by gluconeogenesis and a similar pattern is observed for lactate concentrations in the liver of
newborn rats (Ballard, F.J. (1971) Biochem.J. 124, 265-274).
The delay in lactate reconversion into glucose in the immediately newborn rats has been attributed by Ballard (1971)
to a relative hypoxia as indicated by a decreased state of
adenine nucleotide phophorylation and a highly reduced NAD
redox couple in the liver. Lactate conversion to pyruvate
would be inhibited by the low cytosolic NAD/NADH ratio and
gluconeogenesis would be inhibited by the limited availability of ATP. By 30 minutes after delivery both the ATP/ADP
and NAD/NADH ratios have risen markedly (Ballard, 1971) and
conditions are now favourable for glucose formation from
lactate. At this time the concentrations of lactate in the
blood (Fig.7) and the liver (Ballard, 1971) begin to fall
sharply.

FACTORS AFFECTING INTRA UTERINE GLYCOGEN STORAGE

Jeanne M. Roux

INTRODUCTION

In a review about foetal glycogen reserves, it is obvious to make mention of the work of Claude Bernard (1) who claimed in 1859 that the presence of large amounts of glycogen in the foetal tissues was essential for growth and differentiation.

The specific role of glycogen in the differentiative process remains hypothetic at the present times. Nevertheless, recent quantitative works have shown that the carbohydrate reserves of certain foetal tissues may be of vital importance for both foetus and new born (2). The changes in tissue carbohydrate occuring in accute anoxia and the depletion of the liver glycogen immediately after birth in almost all species (including human) show the importance of this energetic store to maintain the high metabolic rate in the foetus and the new born. Therefore, a defect or a reduction of the glycogen reserves normally present should be a source of clinical problems in the neonatal period.

The purpose of this paper is to evaluate a number of factors which could affect the normal accumulation of glycogen in some organs of the foetus.

THE CARBOHYDRATE RESERVES OF THE FOETUS

Glycogen is present in nearly all the foetal tissues including : brain, kidney, gut, the maximum foetal level being generally higher than the adult one. Accumulation of glycogen in the various tissues occurs at different periods during gestation, with a pattern unique to each tissue.

Shelley (3) has compared the data from different species. It is striking that differences in glycogen levels are relatively low from one specie to the other. Three general patterns have been described depending of the organ. Some organs exhibit a maximum level at mid gestation as in lung or around the three-fourth as in placenta. Other organs : skeletal muscle, liver, accumulate glycogen till birth when maximal values are reached.

The cardiac glycogen follows an inverse pattern, presenting its maximum early in gestation and decreasing continuously till birth.

The species with a relatively long gestation period with the exception of the pig, accumulate glycogen in the liver earlier than those with short gestations, like rat and rabbit. In these cases, the rise occurs in the last fifth of gestation and proceeds at particularly high rate. Since large amounts of glycogen are present in the placenta and lung at a time when there is very little in the liver, it has been suggested that these act as temporary glycogen stores, until the liver is able to take over.

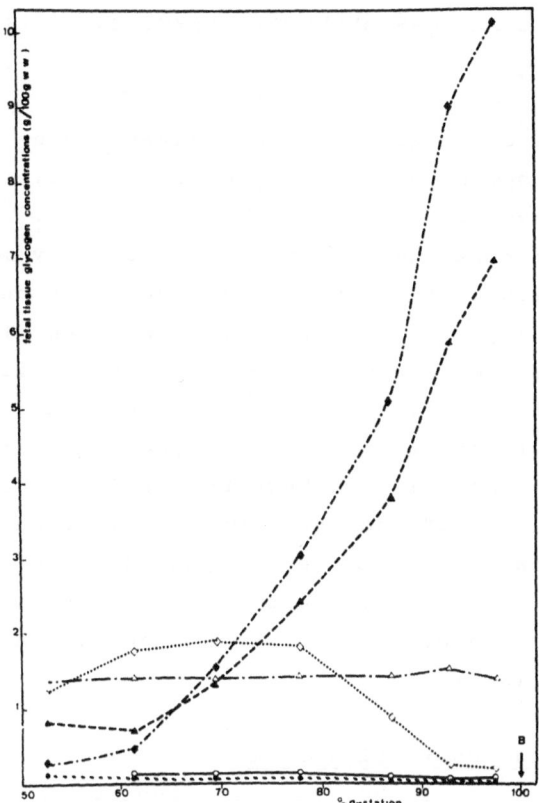

Figure 1. - Glycogen content in some organs of the foetal pig (4).

◆ liver ;   △ heart ;   ◇ lung ;
○ placenta ;   ▲ muscle   ● kidney

The central role of the liver in the regulation of blood sugar levels is well known. For this reason, among the different tissues which accumulate glycogen during the course of the foetal development, the glycogenic storage function of the liver was the most extensively studied. Little is known about the factors responsible for the first appearance of glycogen in the other foetal tissues and for the subsequent changes in its concentration.

FOETAL FACTORS AFFECTING GLYCOGEN STORAGE.

Enzymes.

The appearance of glycogen in a given tissue coincides
with the development of the enzymes responsible for
glycogen synthesis.
It is possible that changes in enzyme activity may
account at least partly for some of the later changes
in glycogen concentration. UDPG glycogen-transglucosilase
appears before the glycogen deposition in the liver (5)
but phosphorylase (6), the degradative enzyme is present
the 14th day of gestation, before glycogen becomes detectable. A direct correlation between the level of enzymes,
quantified "in vitro" and the level of glycogen is
hazardous. Some of the related enzymes have to be activated and the regulation of their activity "in vivo" is
under the strict dependance of both endocrine and metabolic factors (6-7).

Pytuitary-adreno cortical system.

The involvement of the endocrine system in the glycogen
storage was first suggested by Aron (1922) (8) who established a correlation between the organo genesis of the
islets of Langerhans and glycogen deposition in the foetal
liver. Later, Jost and Hatey, 1949 (9) and Jost and
Jacquot, 1955 (10), established the central role of
the pytuitary-adreno cortical system on the glycogen
accumulation in the foetal liver.

In the rat, the glycogen accumulation between days 17
and 18 of gestation, depends on the presence of glucocorticoids in the foetal blood (11). "In vitro", it is
possible to promote glycogen synthesis in foetal
hepatocytes isolated from day 15 foetuses, by culturing
them in the presence of cortisol or dexamethasone (12).
Moreover, glycogen storage can be induced to begin
earlier than during normal development, by injection of

cortisol acetate directly into the 16 day old foetus
"in vivo" (13).

It was recently shown that the synthetic corticoids
betamethasone 17, 21 dipropionate, was not only inactive
to induce glycogen deposition in the foetal liver "in vivo"
and "in vitro", but that it suppressed the hormonal
action of cortisol by a mecanism which remains to be
elucidated (14-15).

In the rabbit, the hormonal regulation followed a more
complicated pattern. If the rabbit foetus was hypophysectomized by decapitation between day 19 and day 24,
further deposition of glycogen was prevented. However,
this phenomenon was not restored by injection of steroids.
A treatment with a saline extract of rat placenta or
purified prolactin or growth hormone associated to
cortisol acetate, allowed the glycogen deposition (10).

In the sheep, glycogen was present in the liver at a
constant level from 100 days of gestation. The rate of
glycogen deposition increased sharply from about
130 days of gestation (16). After hypophysectomy or
adrenalectomy carried out on day 100 or 125 of gestation,
liver glycogen concentration fails to rise (17). Infusion
of cortisol to the foetus "in utero" increases the liver
glycogen in both operated foetuses and the control ones
(18).
Moreover,the glycogen level appeared to be related to
the log. of the plasma corticosteroid concentration (19).
The glycogen deposition can be induced at 115 days and
even as early as 100 days of gestation.
At the present time, it is not established if in this
species, other components act or not in connection with
cortisol.
The glycogen deposition in the foetal liver obeys a dual
hormonal control by corticosteroid and by a pytuitary
hormone which could be supplied by the rat placenta and
probably by the placenta of the ewe.

The effect of the pytuitary-adrenocortical system on the
glycogen storage of the other organ has been studied
by Barnes et al. in the sheep (19). Little or no change
occured in the glycogen concentration, of the skeletal
muscle, heart, lung or placenta, whereas the glycogen
values in perirenal fat are enhanced in adrenalectomized
foetus and depressed after cortisol infusion.
In the rabbit lung, the normal disappearance of glycogen
in the type II pneumocytes of the decapitated foetus
was retarded (20-21). In the pig, decapitation of the
foetus at 43 to 47 days of gestation decreases the glycogen reserve in the longismus muscle at term, but the
authors consider this effect as an artefact (22).

Other Hormones.

Sub-total pancreatectomy "in utero " was perfromed
in 18 day old rat foetus by Felix and Jacquot (23).
Insulin level was decreased by four to five times in
the operated foetuses and liver glycogen content was
slightly reduced. Insulin antibody injections, directly
into the rat foetus "in utero" on day 19 or 21 of
gestation decreased the ^{14}C-glucose incorporation into
the liver glycogen (24).
Neither insulin nor insulin antibody affected the rate
of incorporation in the skeletal muscle. Nevertheless,
a consistent reduction of muscle glycogen level by
exogenous insulin was found. This unexpected action
remained unexplained by the authors (23).
"In vitro" insulin addition stimulates ^{14}C glucose
incorporation into glycogen of foetal hepatocytes in
primary cultures, when grown in the presence of cortisol.
No glycogenic response to insulin was observed in the
culture when cortisol was not previously introduced.
In addition, insulin alone is not able to initiate the
glycogen storage function (25). Thus, insulin completes
the action of the glucocorticoids.
On the other hand, glucagon was shown to induce glycoge-

nolysis in foetal hepatoctyes "in vitro", at all stages studied, even in the absence of cortisol (24). These two peptidic hormones may play a regulatory role on the glycogen metabolism in the foetal liver.

Cardiac stores of glycogen were reduced by glucagon in isolated heart from foetal rats older than 16 days of gestation, whereas it was uneffective in foetal mouse heart, at any period of gestation. Thus, species differences may exist in the responsiveness to hormonal stimuli in addition to the tissue dependant differences previously shown (26).

INFLUENCE OF THE MATERNAL HORMONAL STATUS.

Influence of the maternal hormones on the glycogen deposition in the foetus depends on their placental transfer. Steroids cross readily the placenta whereas the results obtained for both insulin and thyroxine are contradictory (27-28).

Adrenocortical hormones.
Corticosteroids of maternal origin are extensively converted to their biologically inactive 11-dehydro-derivatives, even in the placenta or in the foetal tissues.

The human placenta convert cortisol and prednisolone (respectively 67 % and 51 %) to their 11-ketometabolite whereas the conversion of dexamethasone and betamethasone was low or negligible (29). Michaud and Burton (30) used the ^{14}C corticosterone injections to the pregnant mouse, to clarify the relationships between maternal and fetal corticoids. The data indicated a rapid exchange of hormones in both directions across the placenta. Placenta, head and liver showed distinct different patterns, in the conversion to 11 dehydrocorticosterone which change as a function of time. The existence of an enzyme capable of reducing cortisone to cortisol in the foetal liver and in the foetal lung could induce

age dependant changes in the hormone level of the plasma
and the foetal tissues, regardless of the steroid being
of maternal or foetal origin.
The administration of glucocorticoids directly to the
foetus promote the morphological and biochemical maturation of the lung and the liver (13 - 31 - 32).
It results in a decrease of the glycogen content of the
pneumocyte and an increased level in the liver of
several species : sheep, rat,rabbit.
The maternal administration of glucocorticoids is also
effective, followed by the same effects on glycogen.
Generally, atrophy of adrenals and lower plasma cortisol
concentrations were found in the foetus (33). Moreover,
abnormalities of organ growth and neurologic development
were reported and a controverse has been developed about
the use of prenatal corticosteroids in human pregnancy
(34).
In view of the preceeding observations, it appears that
the corticoids effects on glycogen metabolism and on side-
actions, had to be appreciated with a closer attention
to the choice of the compounds and the schedule of the
administration. The responsiveness of the foetal organs
to corticoids is subject to important changes during
pregnancy.
Epinephrine injected to the pregnant rat has a depleting
effect on the foetal liver glycogen. This effect was
found to be dose dependant (35).

Thyroid and parathyroid hormones.

In the rat, maternal hypothyroidisme has been shown
to decrease the glycogen storage in placenta , skeletal
muscle and liver of the foetuses (35-37).
On the contrary, to the maternal liver which exhibited
a typical metabolism of hypo thyroidism, the foetal liver
presented an enhanced glucose use in "in vitro". The
placenta of the treated animals showed a significantly
greater glucose oxydation to CO_2 and glucose incorporation

into proteins, whereas glucose incorporation into
glycogen and lipids was not affected (38). The decrease
in glycogen content of these animals was not due to
impaired glycogen synthesis, but rather to higher glycogen
turnover and possibly to lower substrate levels. The
"in vivo" foetal liver metabolism, and the elevated
T4 level in the foetal plasma suggested foetal hyper
thyroidism in the foetuses of hypothyroid mothers.
The normal foetal glycogen stores were restored by growth
hormone treatment but only in the presence of sufficient
thyroid hormones (39).
A role of the maternal parathyroid hormones in the
foetal liver glycogen stores is also suggested by the
investigation of Garel and Gilbert in the rat (40).
The glycogen level is depressed after thyroparathyroid-
ectomy of the pregnant rat, and restored by 1,2, 5
dehydroxycholecalciferol.

Maternal diabetes.

The results reported on glycogen level of foetal tissues obtained in diabetic pregnancy are contradictory.
However, all reports are based on autopsy material which
is unsatisfactory for quantitative evaluations (41).
In the placenta and the foetal membrane, the glycogen
level was found to be higher in diabetic than in normal
pregnancy (42). Experimental maternal diabetes induced
in rats by streptozotocin has been also shown to increase
glycogen level in the placenta (43).

In the monkey, the liver from foetuses of streptozotocin treated mothers has increased glycogen levels (44).
When pregnant gilts were subjected to alloxan diabetes,
the total liver glycogen content was significantly elevated in the progeny at 112 days of gestation, whereas
muscle glycogen was only slightly increased (45).

It is well known that the maternal diabetes develops
hyper insulinism in the foetus. The combined effects of
high foetal insulin levels and enhanced glucose supply
promotes the storage of glycogen in the foetal liver.

But, it is important to point out that the principal
effect of maternal diabetes is the increase of cellular
proliferation in foetal tissues, as shown in the pig by
Ezckwe and Martin (45).

Post maturity.

Prolonged pregnancy by injection of progesterone or
synthetic gestagen (46) produce a dramatic depletion in
the liver glycogen stores of the rat foetuses maintained
for two extra days "in utero". Porta et al. (47-48)
related the glycogen fall to a lower insulin level in
the foetal blood and an increased glucose-6-phosphatase
activity.

Acute anoxia.

During acute anoxia, glycogen reserves of the foetus
are depleted except in lung and muscle. An age dependancy
of the carbohydrate mobilization was shown (49). In the
foetal lamb at 90 days of gestational age, anoxia was
initiated by clamping the umbilical cord. After one hour,
glycogen content was unchanged in liver, lung and muscle,
whereas a fall was observed in heart, kidney and brain.
The latter organ was completely depleted. In the foetal
lamb of 121 days of gestation the glycogen store of the
liver was mobilized and blood sugar rose during anoxia
(50). In the foetal rat, glycolysis is more stimulated
by hypoxia in the heart of 16 days of gestation than in
the foetal heart at term (51-52).
In brain, glycogen declined during anoxia, but the energy
consumption was significantly lower in the foetus than in
the neonatal rat,thus glycogen declined more slowly (53).

Nutritional factors.

At the beginning of this century, a number of authors
considered that glycogen content of the foetal tissues,
particularly in the liver, was dependant of the maternal
nutrition, with the exception of the placental reserves

(54) (55). In more recent works, glycogen level is rather
related to the maternal blood glucose.Maternal hypoglycemia produced by insulin injections or prolonged
phlorrhizin treatment (56), produce a depletion in the
liver glycogen whereas placental and foetal muscle glycogen were unaffected.
The concentration of glycogen was not modified in the
liver and the placenta of foetus aged 19 to 21 days
gestation, after fasting of the mother for 16 hours, or
after 2 hours' glucose infusion to the pregnant rat (57).
In the same conditions of fasting, Bassi and Greenberg
(58) found a depletion of the foetal liver glycogen
whereas placental reserves remained unchanged. In spite
of these contradictory results, both groups of authors
concluded after infusion of ^{14}C glucose that foetal
glycogen was in constant equilibrium with fetal blood
glucose.
During hyperglycemia, of the foetus, following glucose
loading of the mother, glycogen synthesis is accelerated
in the foetal liver. These results are age-dependant
as shown by Gilbert and Jost (59). A two hours infusion
of glucose to a pregnant rabbit results in foetal hyperglycemia, but the liver glycogen content doubles only
after 26 days of gestation.

    Girard et al. (60) submitted pregnant rats to prolonged
fasting periods (24 to 96 hours). They showed that maternal fasting induced foetal substrats modifications and
hormonal changes which induced the premature appearance
of hepatic gluconeo genesis. Moreover, glucose could be
spared by the foetus by oxydation of other energetic
sources such as ketone bodies and betahydroxybutyrate
(61) (62).

    The hypothesis of Rossi and Greenberg (58) that a
priviledged situation of the fœtus as regards to the
blood glucose level during fasting could be partly supported by these observations. In the foetal metabolism
priority given to fuel provision to the prejudice of
growth, could be an adaptative mecanism in fasting or

undernutrition.

In view of these recent results, it is interesting to determine to what extent foetal tissue glycogen is affected by undernutrition during gestation.
We studied intra uterine growth retardation induced in the rat by clamping the uterine artery on day 17 of gestation. The reduced blood flow provided via the placenta submitted the foetus to a partial deprivation of nutriments. One day before delivery, the weight of the foetuses in the operated uterine horn, was reduced by 30 % as compared to control (63). All these animals presented a severe hypoglycemia. The hepatic glycogen level was reduced by 40 % as compared to control but cardiac and brain (64) glycogen remained unchanged.

Table I.
Glycogen level and total glycogen content in heart and liver.

	IUGR		CONTROL		NORMAL	
	Glycogen %	Total glyco-gen	Glyco-gen %	Total glyco-gen	Gly-cogen %	Total glyco-gen
Heart (n = 6)	1,83 ± 0,65	0,42	1,91 ± 0,36	0,57	1,42 ± 0,43	0,37
Liver (n = 8).	3,36 ± 1,01	5,00	5,35 ±2,00	19,7	4,63 ± 0,9	14,3

(21 days of gestation), IUGR : intra uterine growth retardation (from Chanez-Bel, 65).

In spite of the restricted blood flow, i.e. metabolites, the foetus accumulated glycogen during late
gestation. The liver is the most stunted organ in
the growth retarded foetus, with a reduction in weight
of more than 60 % of the control (63). Thus, the hepatic
glycogen stores are reduced to one fourth of the reserves
normally present in the foetal liver at term. Nevertheless,
the unchanged heart glycogen level and the presence in
the liver of a glycogen concentration of about 3 % show
that the mecanism of glycogen accumulation is less
impaired than growth.
Glycogen synthesis from glucose did not appear to be
disturbed in I.U.G.R. as shown by Nitzan and Groffman
in foetal rat liver slices (65). We never found in the
IUGR foetuses a premature appearance of the key enzymes
of neoflucogenesis and the total amount of these enzymes
was reduced during a short period after birth (63).
The determination of circulating hormones carried out
by Girard et al. (67) showed a higher level of glucagon
and a lower level of insulin in the IUGR foetuses.
The plasma corticosterone level remained unchanged (68).

In spite of the extensive works on the effect of
undernutrition of the mother on foetal growth, very few
data are found about the foetal glycogen reserves.
In all studies concerning chemical tissue composition,
the results obtained were very similar whether IUGR
was produced by uterine artery clamping, or by undernutrition of the mother. This similarity allows the
assumption that the glycogen metabolism would be modified
at the same extent in both experimental conditions.

CONCLUSION

At the present time, there remains no doubt on the
vital importance of the glycogen reserves accumulated
in the foetal tissues during late gestation. Glycogen
stores in brain, heart, liver and skeletal muscle are
essential to the new born for maintaining intra cellular

ATP, blood glucose and body temperature.
For obvious reasons, there are few informations on the
glycogen reserves of the human foetuses.

Recent "in vitro" studies show that glycogen deposition is under identical hormonal regulation in the human
and in the rat foetal liver (69-70). In the developping
human liver, the concentrations of the various enzymes
tend to change in the same direction as they do in
rat liver (71).

Taking into account the relative length of gestation,
it is reasonable to expect that the factors affecting
glycogen stores in the rat are also effective in the
case of human foetuses, and proceed by the same meca-
nisms.

REFERENCES

1.   Bernard, C, De la matière glycogène de certains
        tissus chez le foetus. C.R. Acad. 8 c. 48 :
        673-684, 1859.

2.   Shelley, H.J., Neonatal hypoglycemia. Br. Med. Bull.
        22 : 34-39, 1966.

3.   Shelley, H.J., Glycogen reserves and their changes
        at birth and in anoxia. Brit. Med. Bull., 17 :
        137-143, 1961.

4.   Randall, G.C., L'ECUYER, R.C.,Tissue glycogen and
        blood glucose and fructose levels in the pig
        foetus during the second half of gestation, Biol.
        Neonate, 28 : 74-82, 1976.

5.   Gilbert, M., Vaillant, R. Contrôle de la synthèse
        du glycogène dans le foie foetal de rat. Biochimie,
        57 : 597-602, 1975.

6.   Coquoin-Carnot, M., et Roux, J.M., Sur les modifica-
        tions du métabolisme glucidique au moment de la
        naissance. J. Physiolgie, 56 : n° 6, 781-806, 1964.

7.   Greengard, O., Enzyme differentiation in mammalian
        liver. Science, 163 : 891-895, 1969.

8.   Arom. Bull. Soc. Chim. Biol., 4 : 209-221, 1922.

9.   Jost, A., and Hatney, J. Influence de la décapita-
        tion sur la teneur en glycogène du foie du foetus
        de lapin, C.R. Soc. Biol., Paris, 143 : 146, 1949.

10.   Jost, A., Jacquot, R., Recherches sur les facteurs
        endocriniens de la charge en glycogène du foie
        foetal chez le lapin avec indications sur le glyco-
        gène placentaire, Ann. Endocrinol., 16 : 849-872,
        1955.

11.   Jacquot, R., Recherches sur le contrôle hormonal
        du glycogène hépatique chez le foetus de rat,
        J. Physiologie, 47 : 857-866, 1955.

12.   Plas, C., Chapeville, F., and Jacquot, R., Develop-
        ment of glycogen storage ability under cortisol
        control in primary cultures of rat fetal hepata-
        cytes, Dev. Biol., 32 : 82-91, 1973.

13.   Greengard, O., Dewey, H.K., The premature deposi-
        tion or lysis of glycogen in livers of fetal rats
        injected with hydrocortisone or glucagon,
        Developmental Biology, 21 : 452-461, 1970.

14.   Mizushima, Y., Ishikawa, M., Hasegawa, Y. Suppression of glycogen storage with betamethazone 17,

21 Dipropionate in fetal rat liver. I. Effect on metabolites of the glucogenic pathway and glycogen synthetase activity., Biochemical Pharmacol., 28 : 737-740, 1979.

15.   Mizushima, Y., Suppression of glycogen storage with betamethasone 17, 21 Dipropionate in fetal rat liver. II. Inhibition of glycogenic action of cortisol and insulin in the liver in organ culture., Biochemical Pharmacol., 28 : 741-745, 1979.

16.   Barnes, R.J., Fowden, A.L., Silver, M., Comline, R.S. Liver glycogen concentrations in the fetal lamb and pig., Ann. Rech. Vet., 8 : 362-373, 1977.

17.   Barnes, R.J., Comline, R.S., Silver, M., The effects of bilateral adrenalectomy or hypophysectomy of the fetal lamb in utero., J. Physiol., 264 : 429-447, 1977.

18.   Barnes, R.J., Comline, R.S., Silver, M., Liver glycogen concentrations in hypophysectomized, adrenalectomized and normal fetal lambs, and the effect of cortisol infusions (proceedings), J. Physiol., 265 : 53-54, 1977.

19.   Barnes, R.J., Comline, R.S., Silver, M., Effect of cortisol on liver glycogen concentrations in hypophysectomized, adrenalectomized and normal fetal lambs during late or prolonged gestation., J. Physiol., 275 : 567-579, 1978.

20.   Meyrick, B., Bearn, J.G., Cobb, A.G., Monkhouse, C.R Reid, L., The effect of in utero decapitation on the morphological and physiological development of the fetal rabbit lung., J. Anat., 119 : 517-535, 1975.

21.   Jost, A., Rieutort, M., Bourdon, J., Plasma growth hormone in the rabbit fetus. Relation to maturation of the liver and lung., C.R. Acad. Sci., 288 : 347-349, 1979.

22.   Kraeling, R.R., Rampacek, G.B., Campion, D.R., and Richardson, R.L., Longissmus   muscle and plasma enzymes and metabolites in fetally decapitated pigs., Growth, 42 : 457-468, 1978.

23.   Felix, J.M., Jacquot, R.L., Effects of sub-total gastro-intestinal pancreasectomy of the rat foetus, J. Endocrinol., 69 : 77-83, 1976.

24. Manns, J.G., Brockman, R.P., The role of insulin in
the synthesis of fetal glycogen., J. Physiol.
Pharmacol., 47 : 917-921, 1969.

25. Plas, C., Nunez, J., Role of cortisol on the glycogenolytic effect of glucagon and on the glycogenic
response to insulin in fetal hepatocyte culture.,
J.B.C., 251 : 1431-1437, 1976.

26. Wildenthal, K., Allen, D.O., Karlson, J., Wakeland,
J.R., Clark, C.M. Jr., Responsiveness to glucagon
in fetal hearts. Species variability and apparent
disparities between changes in beating, adenylase
cyclase activation and cyclic AMP concentration.,
J. Clin. Invest., 57 : 551-558, 1976.

27. Gitlin, D., Kumate, J., and Morales, C., The
transport of insulin across the human placenta.,
Pediatrics, 35 : 65-69, 1965.

28. Dubois, J.D., Clontier, A., Walker, P., Dussault, J.
H., Absence of placental transfer of L-Trisodothyro-
nine int the rat. Pediat. Res., 11 : 116-119, 1967.

29. Blauford, A.T., Pearson Murphy, B.E., In vitro
metabolism of prednisolone, dexamethasone,
betamethasone and cortisol by the human placenta.,
Am. J. Obstet. Gynec. 127 : 264-266, 1976.

30. Michaud, N.J., Burton, A.F., Maternal-fetal
relationships in corticosteroid metabolism.,
Biol. Neonate, 32 : 132-137, 1977.

31. Pines, M., Bashan, N., Moses, S.W., Effect of
hydrocortisone and glucagon on glycogen metabolism
in the fetal rat liver. , Biochem.Biophys. Acta,
411 : 369-376, 1975.

32. Gilden, C., Sevanian, A., Tierney, D.F., Kaplan, S.A
Barrett, C.T., Regulation of fetal lung phosphatidyl
choline synthesis by cortisol : role of glycogen
and glucose., Pediatr. Res., 11 : 845-848, 1977.

33. Epstein, M.F., Farrell, P.M., Sparms, J.W., Pepe, G.
Driscoll, S.G., Chez, R.A., Maternal betamethasone
and fetal growth and development in the monkey.,
Am. J. Obstet. Gynecol., 127 : 261-263, 1977.

34. Frank, L., Roberts, R.J., Effects of low-dose
prenatal corticosteroid administration on the premature rat., Biol. Neonate, 36 : 1-9, 1979.

35.   Delphia, J.M., The glycogen-depleting effect of
        exogenous maternal epinephrine on the liver of
        the fetal rat during the sixteenth day of gestation.
        Res. Commun. Chem. Pathol. Pharmacol. 10 : 569-572,
        1975.

36.   Poterfield, S.P., Hendrich, C.E., The effect of
        maternal hypothyroidism on maternal and fetal tissue
        glucose 1-14 C incorporation in rats. , Horm. Res.,
        6 : 236-246, 1975.

37.   Poterfield, S.P., Whittle, E., Hendrich, C.E.,
        Little, R.C., Hypoglycemia and glycogen deficits
        in fetuses of hypothyroid pregnant rats., Proc. Soc.
        Exp. Biol. Med., 149 : 748-753, 1975.

38.   Poterfield, S.P., The effects of maternal hypothy-
        roidism on the vitro metabolism at   1(-14) C
        glucose in rats., Horm. Metab. Res., 9 : 502-506,
        1977.

39.   Poterfield, S.P., Hendrich, C.E., The effects of
        growth hormone treatment on thyroid-deficient
        pregnant rats on maternal and fetal carbohydrate
        metabolism., Endocrinology, 99 : 786-792, 1976.

40.   Garel, J.M., Gilbert, M., The role of maternal
        parathyroids and vitamin D_3 metabolites in fetal
        growth and the deposition of glycogen reserves in
        the maternal and fetal liver in rats., C.R. Acad.
        Sci., 286 (20) : 1459-1461, 1978.

41.   Farquhar, J.W., Metabolic changes in the infant
        of the diabetic mother. The Pediatric Clinics of
        North America., 12 : 743-764, 1965.

42.   Coquoin-Carnot, M., and Roux, J.M., Sur le métabo-
        lisme du glycogène dans les annexes de l'oeuf
        humain à terme, au cours des grossesses normales
        et diabétiques., Path. Biol., 11 : 21-28, 1963.

43.   Abramovici, A., Sporn, J., Prager, R., Shaltiel, A.,
        Laron, Z., Liban, E., Glycogen metabolism in the
        placenta of streptozotocin diabetic rats.,
        Horm. Metab. Res., 10 : 195-199, 1978.

44.   Glinsmann, W.H., Eisen, H.J., Lynch, A., Chez, R.A.
        Glucose regulation by isolated near-term fetal
        monkey liver., Pediat. Res., 9 : 600-604, 1975.

45.   Ezekwe, Martin, R.J., Influence of maternal alloxan
        diabetes or insulin injections on fetal glycogen
        reserves, muscle and liver development in pigs.,
        J. Anim. Sci., 47 : 1121-1127, 1978.

46. Emmrich, P., Bührdel, P., Willgerodt, H., Morphology of the rat placenta in experimental, hormone-induced prolonged pregnancy. II. Placenta in intra-uterine macerated dead rat foetuses., Endokrinologie, 64 : 205-212, 1975.

47. Portha, B., Rosselin, G., Picon, L., Post maturity in the rat : unpairment of insulin, glucagon, and glycogen stores., Diabetologia, 12 : 429-436, 1976.

48. Portha, B., Le Provost, E., Picon, L., Rosselin, G., Post maturity in the rat : phosphorylase, glucose-6-phosphatose and phosphoenol pyruvete Carboxy-kinase activites in the fetal liver., Horm. Metab. Res., 10 : 141-144, 1978.

49. Dawes, G.S., Mott, J.C., Shelley, H.S., The importance of cardiac glycogen for maintainance of life in foetal lambs and newborn animals during anoxia., J. Physiol., 146 : 516., 1959.

50. Mott,J.C., The ability of young mammals to withstand total oxygen lack. Brit. Med. Bull., 17 : 144-148, 1961.

51. Hoerter, S., Action de l'hypoxie sur la glycolyse du coeur isolé de rat en fin de gestation., C.R. Acad. Sci. /D/- Paris. 286 : 411-414, 1978.

52. Hoerter, J., Evolution de la glycolyse myocardique du rat dans la période périnatale., C.R. Acad. Sci., /D/ Paris. 286 : 465-468, 1978.

53. Vanucci, R.C., Duffy, T.E., Carbohydrate metabolism in fetal and normal rat brain during anoxia and recovery. Am. J. Physiol., 230 (5) : 1263-1275, 1976.

54. Lochhead, T., Cramer, W., The glycogenic changes in the placenta and the fetus of the pregnant rabbit : a contribution to the chemistry of growth., Proc. Roy. Soc. London, 80 : 263-284, 1908.

55. Hugget (A.St.G.)., Maternal control of the placental glycogen., J. Physiol., 67 : 360-371, 1929.

56. Corry, L.E., Growth and glycogen content of the fetal liver and placenta. Am. J. Physiol., 112 : 263-267, 1935.

57. Goodner, C.J., Thompson, D.J., Glucose metabolism in the fetus in utero : the effect of maternal fasting and glucose loading in the rat., Pediat. Res., 1 : 443-451, 1967.

58.   Bossi, E., Greenberg, R.E., Sources of blood glucose
       in the rat fetus.,Pediat. Res., 6 : 765-772, 1972.

59.   Gilbert, M., Jost, A., Hyperglycemia and glycogen
       storage in the rabbit fetal liver. Role of age
       and hormonal status., Biol. Neonate, 32 (3-4) :
       125-131, 1977.

60.   Girard, J.R., Ferre, P., Gilbert, M., Kervran, A.,
       Assan, R., Marliss, E.B., Fetal metabolic response
       to maternal fasting in the rat., Am. J. Physiol.
       232 (5) : E, 456-463, 1977.

61.   Shambaugh, G.E., Mrozak, S.C., Freinkel, N.,
       Fetal fuels. I. Utilisation of ketones by isolated
       tissues at various stages of maturation and maternal
       nutrition during late gestation., Metabolism,
       26 (6) : 623-635, 1977.

62.   Dahlquist, G., Persson, B., Effect of intra-
       uterine growth retardation on the postnatal
       development of D, B, hydroxybutyrate dehydrogenase
       activity in rat brain., Biol. Neonate, 28 :
       353-364, 1976.

63.   Roux, J.M., Chamez-Bel, C., Degremont, C.,
       Gaben-Cogneville, A.M., Fulchignoni-Lataud, M.C.,
       Swierczewski, E., Tordet-Caridroit, C.,
       Minkowski, A., Effect of intra uterine growth
       retardation on cellular proliferation and diffe-
       rentiation in developing rats., Ann. Biol. Anim.
       Bioch. Biophys., 19 (1 B) : 135-150, 1979.

64.   Brown, J.D., Vannucci, R.C., Cerebral oxidative
       metabolism during intra uterine growth retardation.
       Biol. Neonate, 34 : 170-173, 1978.

65.   Chanez-Bel, C., Retard de croissance intra-uterine
       chez le rat., Thèse de l'Université Paris VI,
       Juin 1972.

66.   Mitzan, M., Groffman, H., Glucose metabolism in
       experimental intra uterine growth retardation,
       in vitro studies, with liver and brain slices.,
       Biol. Neonate, 17 : 420-426, 1971.

67.   Girard, J.R., Chanez, C., Kervan, A., Tordet-
       Caridroit, C., Assan, R., Studies on experimental
       hypotrophy in the rat. III. Plasma - insulin -
       and glucagon., Biol. Neonate, 29 : 262-266, 1976.

68.   Chanez-Bel, C., Tordet-Caridroit, C., Influence du
       retard de croissance intra utérine dans le taux de
       corticosterone plasmatique et surrenalien chez le
       rat au cours du développement. C.R. Acad. Sci. ser.
       D., 169 : 286-290, 1975.

69.   Schwartz, Al., Raiha, N.C., Rall, T.W., Hormonal
       regulation of glycogen metabolism in human fetal
       liver. I.   Normal development and effects of
       dibutyryl cyclic AMP, glucagon and insulin in
       liver explants., Diabetes, 24 : 1101-1112, 1975.

70.   Schwartz, Al., Rall, T.W., Hormonal regulation
       of incorporation of alanine -U-14 C into glucose
       in human fetal liver explants. Effect of dibutyryl
       cyclic AMP, glucagon, insulin, and triamcinolone.,
       Diabetes, 24 : 650-657, 1975.

71.   Greengard, O., Enzymic differentiation of human
       liver : comparison with the rat model. Pediat.
       Res., 11 : 669-676, 1977.

================

DISCUSSION

Qu.Salle : What is the content of glycogen in the human placenta, if it is higher than in the rat, what is its role?
An.Roux : The glycogen content of the human and rat placenta are similar. It is generally assumed that this glycogen reserve is utilized only for the **energy** requirements of the placental metabolism.

Qu.Jones : As the effect of IUGR in the monkey and guinea pig on glycogen deposition is different from that in the rat, do you really think that situation in the rat reflect what is happening in IUGR in the human?
An.Roux : I do not think that the situation in either the guinea pig, the monkey or the rat reflects exactly what is happening in the human. A certain number of fundamental biochemical mechanisms are common to all species. The use of various experimental models is necessary to approach the understanding of human hypotrophy.

Com.Shelley : The rabbit is a species where even normal litters contain large and small foetuses. Some years ago, Dr.Paul Harding and I evoked at the carbohydrate reserves in the rabbit foetus and tried to increase the incidence of small foetus by ligating the blood supply to one uterine horn by the Wigglesworth technique. In the small foetuses from normal and "ligated" litters, the hepatic carbohydrate reserves, expressed in terms of concentration per g liver or as the total amount per g body weight, were similar to those of their larger litter mates as in Dr.Jones' guinea pig experiments. But if the mothers's food intake was restricted to 25 % normal,the incidence of small foetuses increased and their liver weight and hepatic carbohydrate concentration/g body weight was half that in their larger litter mates, the situation in Dr.Roux's rats. So in the rabbit, we have seen both phenomena and the difference may be due to the experimental conditions rather than a true species difference. Reference : HARDING,P.G.R. and SHELLEY,H.J.(1966): Some effects of intra-uterine growth retardation in the

foetal rabbit.   In : <u>Intra-uterine dangers to the foetus</u>.
ed. Z.K. Stembera - pp529-531-Excerpta Medica Monograph:
Amsterdam.

<u>An.Roux</u> : Thank you very much Doctor Shelley for your com-
ments concerning the influence of the experimental conditions on the glycogen deposition in the liver of hypotrophic
animals.   I think that many conflicting results depend on
differences in experimental conditions.   The restriction of
the blood flow on the 17th day of gestation in the rat
limits the experiment to the late gestational period.

FACTORS AFFECTING GLUCOSE METABOLISM IN THE NEWBORN RAT

Pascal Ferré, Jean-Paul Pégorier, Soomant Callikan,
Armelle Leturque and Jean Girard

INTRODUCTION

Birth, for most of the mammals is a dramatic change in the nutritional conditions. The fetus receives a continuous transplacental flow of glucose, lactate, amino acids and, according to the species, of free fatty acids (FFA). Glucose is the main oxidative substrate for the fetus although it was shown recently that aminoacids could be important fuels (1). On the other hand, fetal tissues have a poor oxidative capacity for FFA. At birth, fat is the main caloric substrate, whatever the mammalian species (table 1).

Table 1. Milk composition in different mammalian
species expressed in percentage of total
calories

SPECIES	FATS	PROTEINS	CARBOHYDRATES
Human	53	6	41
Monkey	53	12	35
Dog	63	25	12
Pig	59	26	15
Sheep	62	20	18
Guinea pig	62	28	10
Rabbit	58	37	5
Rat	69	25	6

However, glucose-dependent tissues such as brain, red blood cells, intestinal mucosa have still to be supplied with glucose. So, it imposes a completely altered metabolic environment upon the newborn, particularly with respect to glucose homeostasis. It is certainly why hypoglycaemia is of relatively frequent occurrence in human newborns (2). Hypoglycaemia can be induced in the newborn of different species by a short fasting period : pig (3), rabbit (4) and rat (5). In this paper we describe experiments performed in the newborn rats which could contribute to the understanding of mechanisms involved in the development of newborn hypoglycaemia.

GLUCONEOGENESIS AND GLUCOSE HOMEOSTASIS IN THE SUCKLING NEWBORN RAT

The newborn rat receives 69% of its caloric intake as fat and only 6% as carbohydrate from milk(table 1). The glucose intake in the newborn rat, calculated from the milk intake and the lactose concentration of the milk represents only 10% of the total glucose turnover determined in vivo by isotopic methods. Nevertheless, blood glucose is maintained at values around 5 to 6 mM. As the large glycogen stores present in the liver at birth are totally depleted after 12 hours (6), an active glucose production from gluconeogenic precursors (gluconeogenic pathway) should be present in the newborn rat. It is well known that the activities of hepatic gluconeogenic enzymes, low during fetal life, increase rapidly after birth and are elevated during the suckling period (7, 8). The hormonal milieu after birth is extremely favourable for an active gluconeogenesis, with a low plasma insulin and a high plasma glucagon (6). Moreover, gluconeogenic rate determined in vitro in isolated hepatocytes is higher in suckling rats than in fed adults (9, 10). However, the quantitative importance of gluconeogenesis in glucose homeostasis in vivo had never been ascertained. The major role of gluconeogenesis in glucose homeostasis was demonstra-

ted by inhibiting gluconeogenesis <u>in vivo</u>. When 3-mercapto-
picolinate, an inhibitor of phosphoenolpyruvate carboxykinase (key enzyme of the gluconeogenic pathway), was injected in suckling newborn rats , it produced in one hour a
fall in the gluconeogenic rate estimated <u>in vivo</u> using labelled lactate, a decrease in   glycaemia (from 5 to 2
mM) but no change in the high plasma FFA and blood ketone
body concentrations (11). Similar results (12) were obtained using another inhibitor of gluconeogenesis, L-2-amino-
4-methoxy-trans-3-butenoic acid (inhibitor of aspartate
aminotransferase). It is clear from these experiments that
an active gluconeogenesis is essential for glucose homeostasis in the suckling newborn rat, whereas sparing of
glucose by the high concentrations of FFA or ketone bodies
would play only a minor role.

HYPOGLYCAEMIA OF THE STARVED NEWBORN RAT

Whereas suckling newborn rats remain normoglycaemic, the
starved pups, maintained at 37°C and in 80% relative humidity develop a profound hypoglycaemia between 12 and 16h
after birth. An impaired liver glycogenolysis is not responsible for hypoglycaemia as liver glycogen mobilization
follows the same pattern in the suckling and starved newborn rat (6). Rather, it is the low <u>in vivo</u> rate of gluconeogenesis observed in the starved newborn which is the
major cause of hypoglycaemia (13). As emphasized in the
previous section, the development of gluconeogenesis in
the newborn rat depends upon several factors which could
be modified by starvation : a) the development of key gluconeogenic enzymes, b) the hormonal milieu and c) the supply of gluconeogenic precursors and FFA to the liver. The
activities of hepatic phosphoenolpyruvate carboxykinase,
pyruvate carboxylase, fructose 1,6-diphosphatase and glu-
cose-6-phosphatase are identical in both suckling and starved newborns (14). Similarly, the hormonal environment is
appropriate (low insulin and high glucagon) for an efficient gluconeogenesis in starved newborn rats (6). On the

other hand, the comparison of the circulating substrates in
the blood of starved and suckling newborn rats demonstrates
two main differences (figure 1) : the concentrations of
gluconeogenic precursors (lactate, pyruvate, alanine) are
lower in the starved rat ; moreover the concentrations of
fat-derived substrates, FFA, ketone bodies and glycerol are
low.

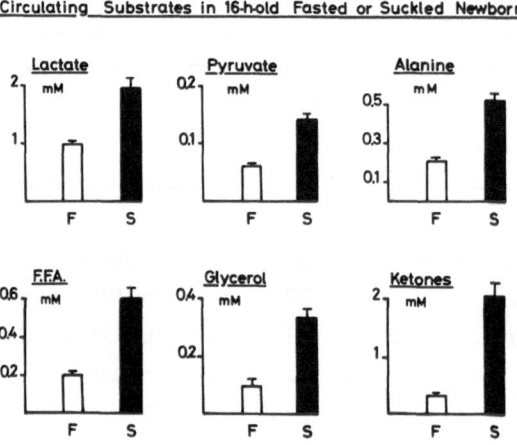

Figure 1 : Circulating substrates in 16h-old
fasted (F) or suckled (S) Newborns

This latter observation is paradoxical in a starved animal;
however, it can be easily explained as the newborn rat is
devoid of fat stores (the small quantities of fat contained
in the brown adipose tissue are not mobilized during starvation at thermoneutrality)(15). In the newborn rat, milk
represents in effect an "external adipose tissue". It was
then hypothetized that a low supply of gluconeogenic substrates and the absence of FFA oxidation in the liver could
impair the gluconeogenic pathway.

CORRECTION OF STARVATION HYPOGLYCAEMIA IN THE NEWBORN RAT

1. External supply of gluconeogenic substrates

In order to increase the low circulating levels of gluconeogenic substrates, newborn rats, starved from birth were
injected at 15h with a mixture of lactate, pyruvate, alanine, serine, glutamine and glycerol.It produced one hour
later a two-fold increase in the glycaemia (16). However,
the glycaemia was still lower than in the suckling rats
(3 mM versus 6 mM), although concentrations of gluconeogenic precursors were higher (16). Thus, deficiency of gluconeogenic substrates cannot explain by itself the impaired gluconeogenesis in the starved newborn rat.

2. External supply of triglycerides

In order to correct the deficiency in fat-derived substrates, newborn rats, starved for 13h were fed with 100 mg
of a triglyceride emulsion or with 100 mg sodium chloride,
using a flexible catheter. In 3 hours, the ingestion of
triglycerides induces a large increase in plasma FFA,
blood ketone bodies and glycerol, indicating that triglycerides have been hydrolyzed and the resulting FFA oxidized in the liver. In these conditions, the glycaemia is
doubled (figure 2) (16).

The rise in blood glucose could be the result of a decreased glucose utilization (due to high plasma FFA and blood
ketone bodies) or an increased glucose production. The
former hypothesis is unlikely as glucose tolerance is not
impaired in triglyceride fed newborns ; moreover, triglyceride feeding is associated with an enhanced plasma insulin
which will tend to increase rather than decrease glucose
utlization. An increased glucose production is not secon-

138

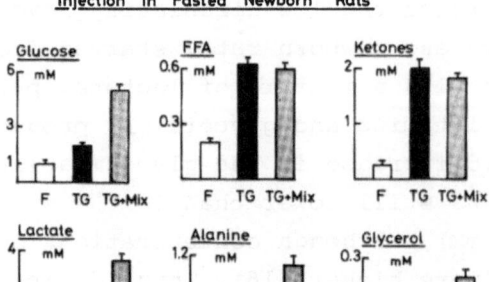

Figure 2 : Fasted newborn rats (F) were fed with
triglycerides (TG) or fed with triglycerides and
injected with a mixture of gluconeogenic substrates
(TG + Mix.)

dary to glycogenolysis as liver glycogen is totally depleted 12h after birth. In fact, several lines of evidence
strongly suggest that triglyceride feeding is associated
with a rise in the gluconeogenic flux : a) an increase in
the conversion of labelled lactate and alanine into glucose
is observed after triglyceride feeding, b) the rise in
blood glucose induced by triglyceride feeding is completely suppressed by the simultaneous injection of 3-mercapto-
picolinate, an inhibitor of gluconeogenesis (16).

3. External supply of both gluconeogenic substrates and
   triglycerides

Despite the increase observed either after gluconeogenic
substrates injection or triglyceride feeding,   glycaemia
in these two experimental conditions is still lower than in
suckling rats (figure 2). If one combines gluconeogenic
substrates injection and triglyceride feeding,   hypoglycaemia induced by starvation is completely reversed in 3
hours (figure 2) (16). Thus fatty acid oxidation has in
vivo a potentiating effect on gluconeogenesis. This conclusion was further validated by inhibiting fatty acid oxidation in the suckling newborn rat using 4-pentenoate. The
inhibition of fatty acid oxidation induces, as it could be
expected, a fall in blood ketone bodies, and is associated
with hypoglycaemia secondary to an inhibition of gluconeogenesis (17).

INTERACTIONS BETWEEN FFA OXIDATION AND
GLUCONEOGENESIS

It is well known that in mammals, acetylCoA derived from
oxidation of FFA cannot give rise to a net glucose synthesis. On the other hand, the possible importance of w-oxida-
tion was recently emphasized in the adult rat (18) ; w -
oxidation of FFA produces succinate which potentially en-
ters the gluconeogenic pathway through oxaloacetate formation (18). However, the quantitative role of this pathway
seems to be minor (10% in diabetic ketoacidosis) (19, 20).
Moreover, it was possible to show in vivo in the starved
newborn rat that at saturating concentrations of gluconeogenic substrates (blood lactate 15 mM) it was still possible to
double the gluconeogenic rate by fat feeding (figure 3).
This renders unlikely an important contribution of the FFA
carbons to glucose through microsomal w -oxidation and
gluconeogenesis.

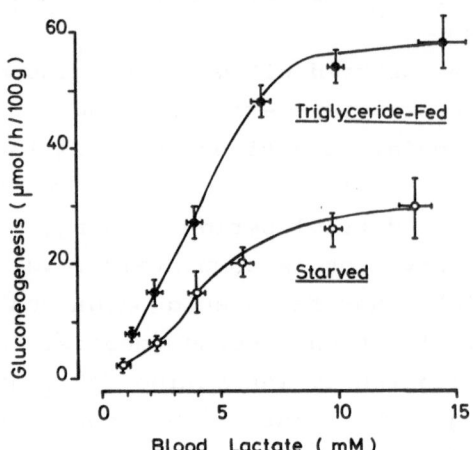

Figure 3 : Gluconeogenesis was estimated using
labelled lactate in 16h-old newborn rats, either
starved from birth or fed with triglyceride and
injected with various amounts of cold lactate to
obtain different blood lactate concentrations.

In recent experiments, we have studied the mechanisms by
which FFA oxidation produces a stimulation of gluconeogenesis (21). To summarize these studies, one can say that an
active fatty acid oxidation stimulates gluconeogenesis by
providing 1) acetylCoA, the obligatory activator of pyruvate carboxylase and 2) the reducing equivalents (NADH)
necessary to displace the reversible reaction catalysed by
glyceraldehyde phosphate dehydrogenase in the direction of
gluconeogenesis.

CONCLUSION

The importance of gluconeogenesis for glucose homeostasis
during the perinatal period is not confined to the rat species. Even in the suckling term human newborns, gluconeogenesis should be an essential process for glucose homeostasis. Taking into account the milk intake of the term human
newborn (70 ml/day/Kg body wt.) and the colostrum composition (5% lactose i.e. 2.5% glucose and 2.5% galactose), one
can estimate the daily glucose intake of the newborn at

about 1.75/day/Kg body wt., during the first 2 days after
birth. When compared to the value of glucose turnover for
newborns of the same age : 8.4 g/day/Kg body wt. (22), it
is clear that an active gluconeogenesis must be present,
even in the breast-fed term newborns, to maintain a normal
glycaemia.

The influence of lipids on glucose homeostasis in the newborn rat cannot be directly extrapolated to other newborns
(and specially to normal human newborns) for a number of
peculiar features of this species : absence of white adipose tissue at birth, primary energy stores in the form
of glycogen and high basal metabolic rate. However, the
small for dates human newborns can be compared in some
ways with newborn rats ; these newborns have limited stores of lipids (white adipose tissue), of carbohydrates
(glycogen) and are subject to frequent neonatal hypoglycaemia. It was possible to show in small for dates that an
infusion of a triglyceride emulsion was associated with an
increased blood glucose which was interprated as resulting
from a stimulation of gluconeogenesis (23, 24), suggesting
that FFA oxidation is necessary for an active gluconeogenesis. So, the informations gathered in the suckling or fasting newborn rats can contribute to the understanding of
mechanisms regulating glucose homeostasis in human newborns.

REFERENCES

1. Battaglia, FC, and G Meschia, Principal substrates of fetal metabolism. Physiol Rev 58 : 499-527, 1978

2. Cornblath, M, and R Schwartz, In : Disorders of Carbo-hydrate Metabolism in Infancy, Philadelphia, Saunders Co, 2nd ed., 1976

3. Swiatek, KR, DM Kipnis, G Mason, K Chao, and M Cornblath, Starvation hypoglycemia in newborn pigs. Amer J Physiol 214 : 400-405, 1968

4. Elphick, MC, The effect of starvation and injury on the utilization of glucose in newborn rabbits. Biol Neonate 17 : 399-409, 1971

5. Girard, JR, GS Cuendet, EB Marliss, A Kervran, M Rieutort, and R Assan, Fuels, hormones and liver metabolism at term and during the early postnatal period in the rat. J Clin Invest 52 : 3190-3200, 1973

6. Girard, JR, P Ferré, A Kervran, JP Pégorier, and R Assan, Influence of insulin:glucagon ratio in the changes of hepatic metabolism during development of the rat, In : Glucagon : Its Role in Physiology and Clinical Medicine, Foa PP, JS Bajaj and NL Foa (eds) Amsterdam, Excerpta Medica, 563-581, 1977

7. Ballard, FJ, and RW Hanson, Phosphoenolpyruvate carbo-xykinase and pyruvate carboxylase in developing rat liver. Biochem J 104 : 866-871, 1967

8. Vernon, RG, and DG Walker, Changes in activity of some enzymes involved in glucose utilization and forma-tion in developing rat liver. Biochem J 106 : 321-329, 1968

9. Beaudry, MA, JL Chiasson, and JH Exton, Gluconeogenesis in the suckling rat. Amer J Physiol 233 : E175-E180, 1977

10. Sly, MR, and DG Walker, A comparison of gluconeogenesis in hepatocytes from neonatal and adult rats. Comp Biochem Physiol 61 : 471-477, 1978

11. Ferré, P, JP Pégorier, and JR Girard, The effects of inhibition of gluconeogenesis in suckling newborn rats. Biochem J 162 : 209-212, 1977

12. Ferré, P., and DH Williamson, Evidence for the partici-pation of aspartate aminotransferase in hepatic glu-cose synthesis in the suckling newborn rat. Biochem J 176 : 335-338, 1978

13. Girard, JR, I Guillet, J Marty, and EB Marliss, Plasma amino acid levels and development of hepatic gluco-neogenesis in the newborn rat. Amer J Physiol 229 : 466-473, 1975

14. Girard, JR, P Ferré, JP Pégorier, A Leturque, and S
    Callikan, Factors involved in the development of
    hypoglycaemia in fasting newborn rats. In : 2nd
    European Symposium on Hypoglycaemia, Andreani, D,
    V Marks, and P Lefebvre (eds), New-York, Academic
    Press, in press, 1980

15. Hull, D, and MJ Hardman, In : Brown Adipose Tissue,
    Lindberg, O (ed.), New-York, Elsevier, 97-115, 1970

16. Ferré, P, JP Pégorier, EB Marliss, and JR Girard, In-
    fluence of exogenous fat and gluconeogenic substra-
    tes on glucose homeostasis in the newborn rat. Amer
    J Physiol 234 : E129-E136, 1978

17. Pégorier, JP, P Ferré, and JR Girard, The effect of
    inhibition of fatty acid oxidation in suckling
    newborn rats. Biochem J 166 : 631-634, 1977

18. Wada, F, and M Usami, Studies on fatty acid ω -oxida-
    tion. Anti-ketogenic effect and gluconeogenicity of
    dicarboxylic acids. Biochim Biophys Acta 487 : 261-
    268, 1977

19. Björkhem, I, On the quantitative importance of ω -oxi-
    dation of fatty acids. J Lipid Res 19 : 585-590,
    1978

20. Kam, W, K Kumaran, and BR Landau, Contribution of ω -
    oxidation to fatty acid oxidation by liver of rat
    and monkey. J Lipid Res 19 : 591-600, 1978

21. Ferré, P, JP Pégorier, DH Williamson, and JR Girard,
    Interactions in vivo between oxidation of non-
    esterified fatty acids and gluconeogenesis in the
    newborn rat. Biochem J 182 : 593-598, 1979

22. Bier, DM, RD Leake, MW Haymond, KJ Arnold, LD Gruenke,
    MA Sperling, and DM Kipnis, Measurement of "True"
    glucose production rates in infancy and childhood
    with 6,6-Dideuteroglucose. Diabetes 26 : 1016-1023,
    1977

23. Mestyan, J, I Rubecz, and G Soltesz, Changes in blood
    glucose, free fatty acids and amino acids in low
    birth weight infants receiving intravenous fat emul-
    sion. Biol Neonate 30 : 74-79,1976

24. Sabel, KG, Metabolic adaptation in small-for-gestatio-
    nal age newborn infants. In : Thesis presented at
    Göteborg University, Department of Pediatrics I,
    University of Göteborg, Göteborg, Sweden, 1978

CARBOHYDRATE METABOLISM IN INTRAUTERINE GROWTH RETARDATION

L.L.A. Kollée, L.A.H. Monnens, J.M.F. Trijbels, J.H. Veerkamp

INTRODUCTION

Intrauterine growth retardation (IUGR) can result from a variety of maternal, placental or fetal factors. Placental insufficiency (24, 25) and chronic vascular disease (36) are of major importance and result in diminution of uterine blood flow. The associated clinical picture is characterized by body wasting, the body weight being low for gestational age. Weight reduction of the organs is the result of diminished cell size (9, 42) and is most pronounced in the liver. The brain/body ratio is elevated because brain weight is relatively spared. The small-for-gestational-age (SGA) infant is threatened by many hazards, one of which is the risk of neonatal hypoglycemia (11). Early glucose administration is, therefore, indicated (2, 11). Onset of hypoglycemia is mostly during the first day after delivery (23, 51). Exogenous glucose is cleared rapidly from the blood in these infants (30). It was accepted for many years that insufficient glycogen stores in SGA newborns could explain their increased risk of neonatal hypoglycemia. Several data, however, point to a more complex mechanism which has not been completely elucidated. The aim of this chapter is to review current knowledge on carbohydrate metabolism in IUGR in rat and man and to present some of our own data in the rat.

EXPERIMENTAL MODEL

In most of the experimental studies on IUGR in the rat the model of Wigglesworth (62) was used. One of both uterine arteries is ligated on the 17th day of gestation. Figure 1 shows a schematic picture.
Near term, at a gestational age of 21 to 22 days, the fetuses are delivered by caesarean section. Birth weight reduction shows a gradient; the fetuses near the ligature are most affected. Several authors confirmed the validity of this model to produce IUGR newborn rats (3, 31,

34, 44, 45, 49, 52, 53, 61).

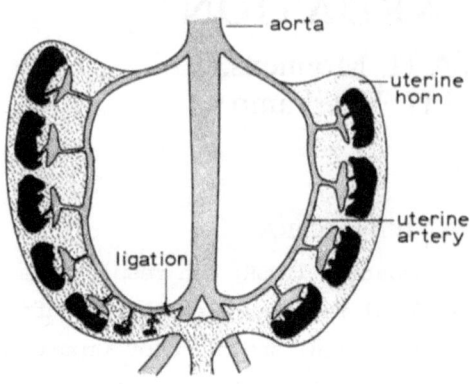

*Figure 1: Schematic picture of the
Wigglesworth model. Modified from (62).*

In the original model, fetuses from the contralateral uterine horns
were used as control animals. In our experiments, we studied normal
animals and animals from sham-operated mothers as well. In the sham-
operated mothers ligation of the uterine artery was omitted. Fetuses
from the contralateral horns are perhaps not the most suitable control
animals for the fetuses from the experimental horns because of effects
of the ligation procedure upon the contralateral horns, as for example
increase of blood-flow.

Table 1 shows mean birth weight of individual animals and mean litter
weight in the 4 groups. The weight in ligated horns is significantly
smaller than that in contralateral horns and litters of sham-operated
dams. The standard deviations in the ligated group and in the group of
animals from contralateral horns are rather large whereas these in the
group of normal rats and animals from sham-operated mothers are smaller.

FETAL CARBOHYDRATE METABOLISM IN IUGR

Reduced uteroplacental blood flow results in a reduction of the supply
of substrates, since the fetus is entirely dependent upon the maternal
circulation. Placental transfer of glucose and aminoacids was recently
investigated by Nitzan and colleagues (46), using labeled 2-deoxyglu-

cose and ∝-aminoisobutyric acid. They found that placental transfer
of these substrates is reduced in term IUGR rats and that the fetal
liver adapts to the reduced glucose supply by enhanced extraction of
glucose from the circulation.

Table 1. *Birthweight of newborn rats after ligation*
of one uterine horn (From (31), with permission)

Birth weight of newborn rats

Group	Birth weight[*] (g) (mean ± SD) (no. of animals in parentheses)	Litter weight[**] (g) (mean ± SD) (no. of litters or horns in parentheses)
Normal litters	4.51 ± 0.39 (473)	4.54 ± 0.28 (47)
Contralateral horns	4.28 ± 0.53 (143)	4.34 ± 0.51 (34)
Ligated horns	3.65 ± 0.59 (104)	3.75 ± 0.49 (28)
Sham litters	4.36 ± 0.37 (185)	4.38 ± 0.27 (21)

[*]Of individual animals
[**]Calculated from mean weights of individuals within litters or horns

Blood glucose levels in IUGR rat fetuses are lower than in control fetuses (5, 43, 52). As a result of diminished glucose supply, hepatic
glycogen deposition is reduced (37, 43, 48, 52), especially during the
last 2 days of gestation. The liver of the IUGR fetus seems, however,
to be enzymatically well equipped for glycogen synthesis (44, 47). A
significant correlation between mean plasma glucose levels and total
liver glycogen content at birth was established in IUGR rats (37).
Glucose infusion into the mother prior to delivery results in a higher
blood glucose level and hepatic glycogen content at birth in these
rats (29, 48).

Some hormonal changes were demonstrated by Girard and his colleagues
to occur in the IUGR rat fetus near term (20, 22). Plasma insulin and
glucagon show lower and higher values, respectively, in these fetuses
compared to controls. Corticosterone levels in fetal IUGR rats were
similar to those in controls (8).

CARBOHYDRATE METABOLISM IN THE EARLY NEONATAL PERIOD IN IUGR

Glucose levels in this period in IUGR rats (6, 7, 31, 37, 50, 56) and
human babies (11) are lower than those in controls. This is illustrated
for the rat in Figure 2. Blood glucose concentrations show the well-
known pattern of a decrease during the first hours after birth, followed by an increase (4, 19, 57). The transient hypoglycemia in the first
hours after birth reflects the change-over to glycogenolysis and gluconeogenesis. The low blood glucose concentration at birth and 1 h after birth in animals from litters of sham-operated dams compared to
normal animals indicates an effect of the operation procedure, including anesthesia, on carbohydrate metabolism during gestation. Levitsky
(35) found no difference in blood glucose values at birth, but a lower
glucose value at 4 h of age in animals from sham-operated dams compared
to normal newborn rats. The finding of higher glucose values at birth
in animals from contralateral horns compared to those from sham-litters
indicate some effect of the ligation procedure on glucose metabolism in
fetuses from the contralateral horns. From 3 h after delivery, blood
glucose values in our animals from sham litters rise significantly more
rapid than in animals from ligated horns. Animals from ligated horns
show significantly lower blood glucose levels than animals from normal
litters and from contralateral horns.

MEAN GLUCOSE LEVELS

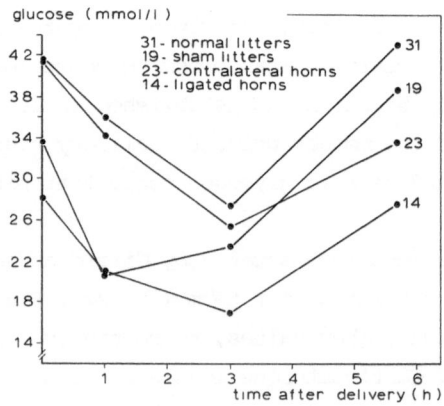

*Figure 2: Blood glucose levels in the early neonatal period
in IUGR (31).*

Glycogen stores are rapidly mobilized after birth. The rate of glycogen break-down in IUGR newborn rats is the same as in animals from contralateral horns (6). The largest part of hepatic glycogen stores is mobilized within 6 h after birth and after 24 h hepatic glycogen content reaches its lowest value in both groups.

Insulin and glucagon are important hormones in the regulation of carbohydrate metabolism in the neonatal period (21). Snell suggested that the rise in plasma glucagon after birth initiates gluconeogenesis and that the fall in plasma insulin triggers glycogenolysis (58). In both the IUGR and normal rat, glucagon levels increase and insulin levels decrease after birth as shown by Girard and his colleagues (22). The increase of blood glucose levels after administration of glucagon to hypoglycemic IUGR newborn babies during the first day of life is not different from that in nonhypoglycemic SGA and normal newborns (14). No differences in plasma insulin and glucagon levels were found in the early neonatal period between SGA and normal newborn babies (60). These data indicate that neonatal hypoglycemia in IUGR may not be due to inappropriate secretion of these hormones.

Hepatic gluconeogenesis becomes active after birth and is thought to contribute to the increase of blood glucose levels after about 3 h of age in the rat (21). Gluconeogenesis was not extensively studied in IUGR. In liver slices of IUGR rats, the conversion of labeled alanine to glucose is lower compared to control animals (44). The observed effects of administration of alanine on glucose levels in newborn babies (17, 39, 59, 63) are confusing and difficult to interpret. Stimulation of glycogen breakdown by alanine which provokes glucagon secretion and utilization of glucose have to be taken into account.

Gluconeogenic key enzyme activities increase shortly after birth in the normal rat (21). With regard to the development of these enzymes in IUGR no data are available on the human infant but some data were published on the rat (6, 32, 50, 56). Development of glucose-6-phosphatase activity after birth in IUGR rats is similar as in rats from contralateral horns (6, 32, 56).

When IUGR animals are compared with animals from sham-operated dams, however, we found a lower activity of this enzyme (32) (Figure 3). A slight delay in the development of fructose-1,6-diphosphatase activity was found in IUGR rats compared to animals from contralateral horns (6, 32, 56).

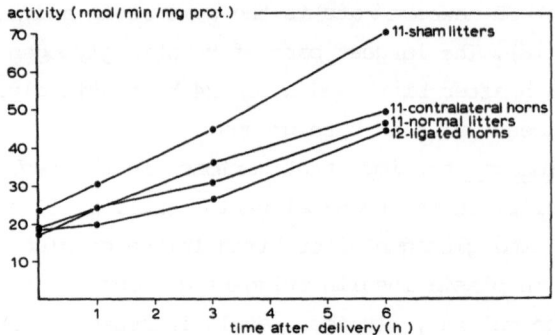

ACTIVITY OF GLUCOSE -6 - PHOSPHATASE
IN NEW BORN RAT LIVER

Figure 3: *Activity of glucose-6-phosphatase in newborn rat liver*

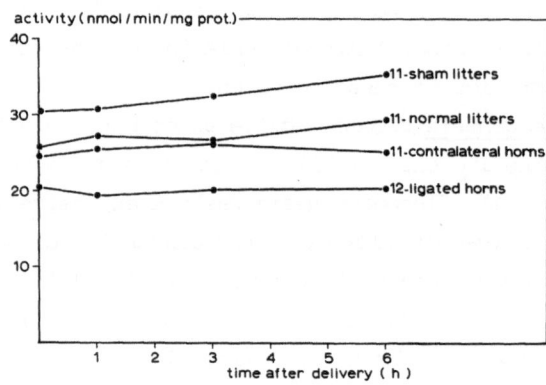

ACTIVITY OF FRUCTOSE-1,6- DIPHOSPHATASE
IN NEW BORN RAT LIVER

Figure 4: *Activity of fructose-1,6-diphosphatase in new-born rat liver*

In our experiments, we also observed a lower activity compared to animals from sham litters (32) (Figure 4). Phosphoenolpyruvate carboxykinase activity in IUGR rats was reported recently by Pollak and colleagues (50), who compared IUGR rats with animals from sham litters. They found a significantly lower activity of this enzyme at 4 h after birth. In our own experiments (32) (Figure 5) we found that the activity of this enzyme increased more rapid from 3 h after birth in newborns from sham litters than in animals of the other 3 groups. Pyruvate carboxylase activities were significantly lower until 6 h after birth in IUGR rats compared to animals from sham litters (32) (Figure 6). The available data indicate that the development of gluconeogenic key enzymes in IUGR is delayed to some extent.

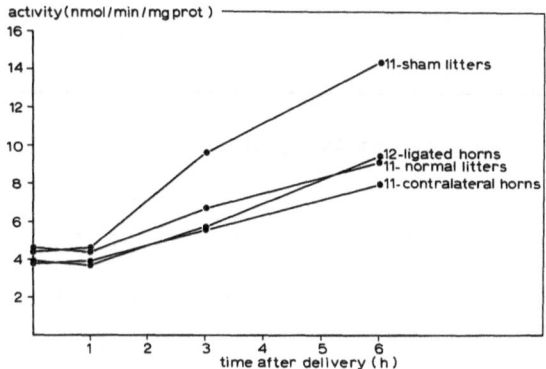

Figure 5: Activity of phosphoenolpyruvate carboxykinase in newborn rat liver

Some data on gluconeogenic substrates will be reviewed next. Of them, amino acids are important. A few studies on amino acid metabolism in IUGR have been performed (28, 38, 40, 54). Significantly higher total blood amino acid levels were found at birth in IUGR rats compared to control animals. The gluconeogenic amino acids alanine, glycine, proline and valine were particularly elevated, but decreased rapidly from 2 - 4 h after birth (38). We could confirm that most amino acid concentrations are elevated in IUGR rats compared to animals from sham litters.

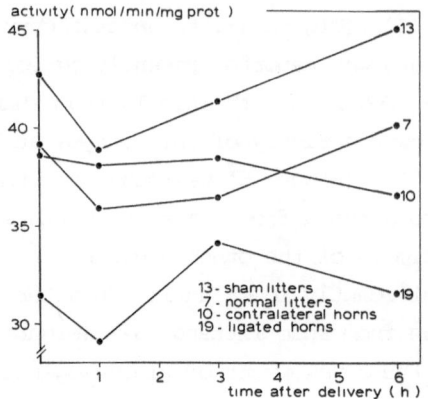

ACTIVITY OF PYRUVATE CARBOXYLASE
IN NEWBORN RAT LIVER

Figure 6: *Activity of pyruvate carboxylase in newborn rat liver*

Table 2. *Some plasma amino acids in IUGR and sham-control newborn rats*

Amino acid		Time after delivery (h)			
		0	1	3	6
Alanine					
IUGR	(10)	1317 ± 273	1176 ± 513	963 ± 914	540 ± 243
Sham	(5)	1051 ± 145	876 ± 268	536 ± 287	442 ± 143
Glycine					
IUGR	(10)	944 ± 182	1087 ± 256	926 ± 303	1061 ± 235
Sham	(5)	555 ± 93	475 ± 78	491 ± 66	565 ± 55
Threonine					
IUGR	(10)	298 ± 38	240 ± 108	157 ± 77	157 ± 30
Sham	(5)	231 ± 75	133 ± 53	90 ± 18	94 ± 15

Values are given as μmol/l as the mean ± SD of the number of litters or horns between parentheses.

Some of them are shown in Table 2. In SGA newborn babies amino acid levels are also higher than in normal newborns (28, 40). This hyperaminoacidemia is even more pronounced in hypoglycemic SGA newborns (40). The
IUGR individual appears to accumulate important gluconeogenic substrates during the first period after birth.

Gluconeogenesis is stimulated by fatty acid oxidation which increases
the intramitochondrial concentration of acetyl-CoA and NADH (15).
Ferré has shown that neonatal hypoglycemia in the fasted newborn rat
can partially be reversed by oral feedings with triglycerides or gluconeogenic substrates (15).

FFA levels normally rise after birth, as shown by Ferré (16), but in
IUGR newborn rats they were lower than in animals from contralateral
horns at birth and during the first day of life (7). We measured FFA
levels in IUGR rats as well (Table 3) and confirm that they are lower
than those in animals from contralateral horns during the first 6 h after birth. Compared with animals from sham litters, however, the IUGR
rat had elevated levels of FFA.

We have no explanation for this finding. Reports of FFA levels in SGA
newborn infants (1, 10, 18, 26, 60) are confusing.

A smaller increase of FFA levels from birth to 6 h of age was found in
hypoglycemic SGA newborns when they were compared to nonhypoglycemic
SGA controls (33). When low-birth-weight infants are fed with fat emulsions the rates of fatty acid oxidation and gluconeogenesis increase
(41). Ferré (15) suggested that a lack of availability or oxidation of
fatty acids in IUGR may result in insufficient gluconeogenesis.

Preliminary data from our laboratory do not show differences in hepatic
fatty acid oxidation between IUGR and control rats (Table 4).

Some attention must be payed to the development of ketogenesis because
ketone bodies can be used as alternative substrates by the brain (27).
No differences in plasma concentration of ketone bodies between IUGR
and control rats could be detected (12). A decreased cerebral uptake of
ketone bodies, however, was measured at 20 days of age in fasting IUGR
rats (13). In SGA newborn infants, ketone body concentrations remained
low from 2 h after birth in contrast to the increase which was observed
in normal infants (28).

In a more recent study, no differences in ketone body levels were found
in the early neonatal period between SGA and normal infants (60).

Table 3. Free fatty acid levels in the early neonatal period in IUGR

Group		Time after delivery (h)			
		0	1	3	6
Normal litters	(5)	0.23 ± 0.05	0.26 ± 0.06	0.36 ± 0.02	0.37 ± 0.07
Contralateral horns	(5)	1.08 ± 0.16	0.89 ± 0.10	1.03 ± 0.18	1.03 ± 0.10
Ligated horns	(5)	0.43 ± 0.18	0.43 ± 0.06	0.45 ± 0.14	0.42 ± 0.12
Sham litters	(5)	0.16 ± 0.08	0.16 ± 0.07	0.24 ± 0.04	0.19 ± 0.04

Values are given in mmol/l as the mean ± SD of the number of litters or horns between parentheses.

Administration of fat emulsions to SGA newborns at 4 h of age results
in an increase of the B -hydroxybutyrate concentration (55).

Table 4. Fatty acid oxidation rate in liver homogenates of newborn and adult rats

	nmol palmitate/min per mg protein
Normal litters	0.16 ± 0.03 (4)
Contralateral horns	0.20 ± 0.05 (6)
Ligated horns	0.19 ± 0.06 (4)
Sham litters	0.16 ± 0.04 (4)
Adult animals	1.11 ± 0.24 (7)

Values are means $\pm$ SD of the number of experiments given in parentheses
and were determined as nmol palmitate oxidized to $^{14}CO_2$ and ^{14}C-labeled
acid-soluble products. Age of newborn animals was 1 h.

CONCLUSIONS

The following conclusions can be drawn from the available data:

1. Animals from sham-operated mothers are to be preferred as controls in experimental IUGR by ligation of the uterine artery in the rat.
2. Less glycogen is stored in the fetal liver due to reduced glucose supply in IUGR.
3. Plasma glucose levels are decreased during the early neonatal period after IUGR.
4. Development of gluconeogenic capacity after birth in IUGR might be delayed because gluconeogenic key enzyme activities are lower compared to controls in the early neonatal period. The accumulation of gluconeogenic substrates can be explained by this delay in development.
5. It remains to be established whether a disturbed fat metabolism in IUGR influences gluconeogenesis.
6. Hormonal regulation by insulin and glucagon of carbohydrate metabolism in IUGR appears to be intact.
7. Increased glucose utilization, for example by the brain, might contribute to the development of neonatal hypoglycemia in IUGR.

REFERENCES

1. Anagnostakis, D and R Lardinois, Urinary catecholamine excretion and plasma NEFA concentration in small for date infants. Pediatrics 47:1000-1009,1971
2. Bossi, E, Neonatale hypoglykämie. Eine Übersicht über pathophysiologie, Klinik und Therapie. Schweiz Rundschau Med 64:1214-1219, 1975
3. Brans, YW and P Ortega, Water content and distribution in intra-uterine growth-retarded newborn rats. Biol Neonate 31:116-121, 1977
4. Cake, MH, D Yeung, and IT Oliver, The control of postnatal hypoglycemia. Suggestions based on experimental observations in neonatal rats. Biol Neonate 18:183-192,1971
5. Chanez, C, JM Roux, and C Tordet-Caridroit, Glycémie, glycogène et glucose-6-phosphatase dans la foie, à la période périnatale,chez le rat dysmature. C R Soc Biol 163:2272-2275,1969
6. Chanez, C, C Tordet-Caridroit, and JM Roux, Studies on experimental hypotrophy in the rat. II. Development of some liver enzymes of gluconeogenesis. Biol Neonate 18:58-65,1971
7. Chanez, C and C Tordet-Caridroit, Glucose, acides gras libres et glycerol du plasma, au cours du développement au rat ayant subi un retard de croissance intra-utérine. Arch Franc Péd 29:593-601, 1972
8. Chanez-Bel, C and C Tordet-Caridroit, Influence du retard de croissance intra-utérine sur le taux de corticostérone plasmatique et surrénalien chez le rat au cours du développement. C R Soc Biol 169:286-290,1975
9. Chase, HP, CS Dabiere, N Welch, and D O'Brien, Intra-uterine undernutrition and brain development. Pediatrics 47:491-500,1971
10. Christensen, NC, Concentrations of triglycerides, free fatty acids and glycerol, in cord blood of newborn infants with a birth weight of 2700 grams. Acta Paediat Scand 66:43-48,1977
11. Cornblath, M and R Schwartz, Hypoglycemia in the neonate,In: Disorders of carbohydrate metabolism in infancy, Ch 5, Saunders, Philadelphia,1976
12. Dahlquist, G and B Persson, Effect of intrauterine growth retardation on the postnatal development of D-B-hydroxybutyrate dehydrogenase activity in rat brain. Biol Neonate 28:353-364,1976
13. Dahlquist, G, Cerebral utilization of glucose, ketone bodies and oxygen in starving infant rats and the effects of intrauterine growth retardation. Acta Physiol Scand 98:237-247,1976
14. Falorni, A, F Massi-Benedetti, S Gallo, and A Romizi, Levels of glucose in blood and insulin in plasma and glucagon response to arginine infusion in low birth weight infants. Pediat Res 9:55-60,1975
15. Ferré, P, JP Pégorier, EB Marliss, and JR Girard, Influence of exogenous fat and gluconeogenic substrates on glucose homeostasis in the newborn rat. Amer J Physiol 234:129-136,1978a
16. Ferré, P, JP Pégorier, DH Williamson, and JR Girard, The development of ketogenesis at birth in the rat. Biochem J 176:759-765, 1978b
17. Frazer, TE, IE Karl, L Hillman, and DM Bier, Direct measurement of gluconeogenesis from 2,3 - ^{13}C alanine in the human neonate during the first 8 hours of life. Pediat Res 13:474 (abstract), 1979

18. Gentz, JCH, R Warrner, BEH Persson, and M Cornblath, Intravenous glucose tolerance, plasma insulin, free fatty acids and B-hydro-xybutyrate in underweight newborn infants. Acta Paediat Scand 58:481-490,1969

19. Girard, JR, GS Cuendet, EB Marliss, A Kervran, M Rieutort, and R Assan, Fuels, hormones, and liver metabolism at term and during the early neonatal period in the rat. J Clin Invest 52: 3190-3200,1973

20. Girard, JR, A Kervran, E Soufflet, and R Assan, Factors affecting the secretion of insulin and glucagon by the rat fetus. Diabetes 23:310-317,1974

21. Girard, J, P Ferré, and M Gilbert, Le metabolisme énergétique pendant la période périnatale. Diab Metab (Paris) 1:241-257,1975

22. Girard, JR, C Chanez, A Kervran, C Tordet-Caridroit, and R Assan, Studies on experimental hypotrophy in the rat. III. Plasma insulin and glucagon. Biol Neonate 29:262-266,1976

23. Griffith, AD, Association of hypoglycaemia with symptoms in the newborn. Arch Dis Child 43:688-694,1968

24. Gruenwald, P, Chronic fetal distress and placental insufficiency. Biol Neonate 5:215-265,1963

25. Gruenwald, P, Chronic fetal distress. Clin Ped 3:141-149,1964

26. Harris, RJ, Plasma nonesterified fatty acid and blood glucose levels in healthy and hypoxemic newborn infants. J Pediat 84:578-584,1974

27. Hawkins, RA, DH Williamson, and HA Krebs, Ketone-body utilization by adult and suckling rat brain in vivo. Biochem J 122:13-18, 1971

28. Haymond, MW, IE Karl, and AS Pagliara, Increased gluconeogenic substrates in the small-for-gestational-age infant. New Engl J Med 291:322-328,1974

29. Hohenauer, L, Studien zur intrauterinen Dystrophie. I. Intrauterine Dystrophie im Tierexperiment. Pädiat Pädol 6:1-16,1971

30. Horváth, I, P Tóth, and K Méhes, The predictive value of glucose utilization rate in neonatal hypoglycaemia of small-for-gestational-age infants. Acta Paediat Acad Sci Hung 16:143-147,1975

31. Kollée, LAA, IAH Monnens, JMF Trijbels, JH Veerkamp, and AJM Janssen, Experimental intrauterine growth retardation in the rat. Evaluation of the Wigglesworth model. Early Hum Dev 3:295-300, 1979

32. Kollée, LAA, IAH Monnens, JMF Trijbels, JH Veerkamp, AJM Janssen, and H van Haard-Hustings, Gluconeogenic key enzymes in normal and intrauterine growth-retarded newborn rats. Early Hum Dev 3: 345-352,1979b

33. de Leeuw, R and IJ de Vries, Hypoglycemia in small-for-dates newborn infants. Pediatrics 58:18-22,1976

34. Levitsky, LL, SM Speck, and R Shulman, Metabolic response to fasting in experimental intrauterine growth retardation. A comparison of two models. Biol Neonate 30:11-16,1976

35. Levitsky, LL, A Kimber, JA Marchichow, and J Uchara, Metabolic response to fasting in experimental growth retardation induced by surgical and nonsurgical maternal stress. Biol Neonate 31: 311-315,1977

36. Lubchenco, LO, C Hansman, and L Bäckström, Factors influencing fetal growth, In: Aspects of praematurity and dysmaturity, Jonxis, JHP, HKA Visser, and JA Troelstra(ed.), Leiden, Stenfert Kroese, 149-166,1968

158

37. Manniello, RL, AJ Adams, and PM Farrell, The influence of antena-
    tal corticosteroids on hypoglycemia in newborn rats with intra-
    uterine growth retardation. Pediat Res 11:840-844,1977a
38. Manniello, RJ, JD Schulman, and PM Farrell, Amino acid metabolism
    in dysmature newborn rats - possible explanation for the anti-
    hypoglycemic effects of prenatal glucocorticoids. Pediat Res 11:
    1165-1166,1977b
39. Mestyán, J, K Schultz, and M Horváth, Comparative glycemic respon-
    ses to alanine in normal term and small-for-gestational-age in-
    fants. J Pediat 85:276-278,1974
40. Mestyán, J, Gy Soltész, K Schultz, and M Horváth, Hyperaminoacide-
    mia due to the accumulation of gluconeogenic amino acid precur-
    sors in hypoglycemic small-for-gestational-age infants. J Pediat
    87:409-414,1975
41. Mestyán, J, I Rubecz, and Gy Soltész, Changes in blood glucose,
    free fatty acids and amino acids in low birth-weight infants re-
    ceiving intravenous fat emulsion. Biol Neonate 30:74-79,1976
42. Naye, RL and JA Kelly, Judgment of fetal age. III. The patholo-
    gist's evaluation. Pediat Clin N Amer 13:849-862,1966
43. Nitzan, M and H Groffman, Metabolic changes in experimental intra-
    uterine growth retardation: blood glucose and liver glycogen in
    dysmature and premature newborn rats. Israel J Med Sci 6:697-
    702,1970
44. Nitzan, M and H Groffman, Glucose metabolism in experimental in-
    trauterine growth retardation.In vitro studies with liver and
    brain slices. Biol Neonate 17:420-426,1971a
45. Nitzan, M and H Groffman, Hepatic gluconeogenesis and lipogenesis
    in experimental intrauterine growth retardation in the rat. Amer
    J Obstet Gynec 109:623-627,1971b
46. Nitzan, M, S Orloff, and JD Schulman, Placental transfer of ana-
    logs of glucose and amino acids in experimental intrauterine
    growth retardation. Pediat Res 13:100-103,1979
47. Oh, W, M D'Amodio, LL Yap, L Hohenauer, and J Metcoff, Glycogen
    synthesis in experimental intrauterine fetal growth retardation.
    Pediat Res 2:415-416,1968
48. Oh, W, M D'Amodio, LL Yap, L Hohenauer, and J Guy, Carbohydrate
    metabolism in experimental intrauterine growth retardation in
    rats. Amer J Obstet Gynec 108:415-421,1970
49. Oh, W and JA Guy, Cellular growth in experimental intrauterine
    growth retardation in rats. J Nutr 101:1631-1634,1971
50. Pollak, A, JB Susa, BS Stonestreet, R Schwartz, and W Oh, Phospho-
    enolpyruvate carboxykinase in experimental intrauterine growth
    retardation in rats. Pediat Res 13:175-177,1979
51. Raivio, KO, Neonatal hypoglycemia. II. A clinical study of 44
    idiopathic cases with special reference to corticosteroid treat-
    ment. Acta Paediat Scand 57:540-546,1968
52. Roux, JM, C Tordet-Caridroit, and C Chanez, Studies on experimen-
    tal hypotrophy in the rat. I. Chemical composition of the total
    body and some organs in the rat foetus. Biol Neonate 15:342-347,
    1970
53. Roux, JM, Studies on cellular development in the suckling rat with
    intrauterine growth retardation. Biol Neonate 18:290-299,1971
54. Roux, JM and Th Jahchan, Plasma level of amino-acids in the de-
    veloping young rat after intra-uterine growth retardation. Life
    sci 14:1101-1107,1974
55. Sabel, KG, R Olegard, K Hildingsson, M Mellander, and P Karlberg,

Imparied fatty acid oxidation and increased gluconeogenic plasma substrates in SGA newborns with hypoglycemia-improvement after injection of lipids. Pediat Res 13:72 (abstract),1979

56. Siegel, SR, W Oh, and DA Fisher, Fructose-1,6-diphosphatase and glucose-6-phosphatase in newborn rats with intrauterine growth retardation. Early Hum Dev 3:43-49,1979

57. Snell, K and DG Walker, Glucose metabolism in the newborn rat. Temporal studies in vivo. Biochem J 132:739-752,1973

58. Snell, K and DG Walker, Glucose metabolism in the newborn rat: the role of insulin. Diabetologia 14:59-64,1978

59. Søvik, O and PH Finne, Alanine-stimulated glucose production in the small-for-gestational-age infant. Pediat Res 11:1024 (abstract),1977

60. Stanley, ChA, EK Anday, L Baker, and M Delivoria-Papadopoulos, Metabolic fuel and hormone responses to fasting in newborn infants Pediatrics 64:613-619,1979

61. Tordet-Caridroit, C, J Roux, and C Chanez, Etude du développement post-natal du rat né dysmature. C R Soc Biol 163:1321-1323,1969

DISCUSSION

<u>Qu.Jones</u> : Is the interpretation of your difference between contralateral horn-fetuses and those that are sham-operated that uterine artery ligation at day 17 in the rat has effects on fetal development that cannot be oxidized to reduction of uterine blood flow related.

<u>An.Kollee</u> : From the results of birth weight and blood glucose measurements we conclude that ligation of one uterine artery at day 17 has effects on growth and metabolism of the fetuses in the opposite horn.

<u>Com.Minkowski</u>: I am interested in the high level of amino acids because that could be potentially dangerous for the liver.

<u>An.Kollee</u> : From the fact that amino acid levels in the fetus before birth are elevated and higher than those in the maternal circulation it might be concluded that elevation of amino acid levels may be tolerated by the liver in the early neonatal period as well.

<u>Qu.Hull</u> : Have you measured the fat content of the liver of the growth retarded rats? There is evidence that in guinea pigs and rabbits and possibly also in some instances in humans, the livers of undersized fetuses at term have high fat contents.

<u>An.Kollee</u> : We did not measure the fat content of the livers and are not aware of any data on hepatic fat in IUGR in the rat. Total body fat content, however, is decreased.

<u>Qu.Jones</u> : How are the changes you reported influenced by the extent of reductionin fetal size? In the fetal guinea pig the functional effects of IUGR are very dependent on the extent of the reduction of fetal size.

<u>An.Kollee</u> : We did not calculate our results in relation to the extent of growth retardation, as you did yourself. However, we noticed that a direct relationship between birth weight and blood glucose levels does not exist.

Qu.Milner : Complimented in experimental design- sham-operated controls versus normal is a second experiment grafted
into ligated versus sham-operated.
Pointed out that human small for dates may be normoinsuli-
naemic or hyperinsulinaemic (La Dune,Arch.Dis.Child,1965)
and that human small for dates are heterogenous.
An.Kollee : The experimental model of reduction of uterine
blood flow reflects, of course, only one of the pathogenic
mechanisms leading to IUGR. It is generally accepted, however, that chronic vascular disease of the mother is of
major importance in the pathogenesis of IUGR in the human
fetus.
No differences in plasma insulin levels were found in the
early neonatal period between small-for-gestational-age and
normal newborn babies (Stanley et al.,Pediatrics, 64, 613-
619, 1979).

Qu.Girard : How do you explain the high plasma FFA levels
observed in newborns from contralateral horns? It is
difficult to explain if we accept that FFA transfers across
the placenta are very low in the rat.
An.Kollee : We were surprised by the high plasma FFA levels
in newborn rats from contralateral horns, but have no explanation for this. Dr.Roux observed the same phenomenon
(personal communication). FFA concentrations were determined using 63 Ni as tracer according to Ho (Anal.Biochem.36;
105-113,1970), which method is generally accepted as a
reliable one.

Qu.Van Assche : A problem of the Wigglesworth technique is
the irregular distribution of IUGR. Should it not be
better to use a model with equal distribution of IUGR?
An.Kollee : A model in which IUGR is distributed equally is
not available. However, such a model would not reflect the
biological variation which occurs in human IUGR.

Com.Roux : In intrauterine growth retardation, if the ligation is made on the 16th day of gestation, most of the foe-

tuses of the ligated horn are resorbed. We observed no
differences between the results obtained when ligation was
done on the 17th or 18th day of gestation.
An.Kollee : Fetal growth in the rat is enhanced during the
last four days of gestation starting between the 17th and
18th day. Ligation during this period results in IUGR.
We also observed, in preliminary experiments, that ligation
before this period kills many of the still immature fetuses.

Qu.Sauer : In growth retarded infants who are hypoxic after
birth, we observe a high glucose level. Do you have any
explanation for this? It shows that these infants have the
possibility to raise their glucose level.
An.Kollee : Hypoxia may result in increased utilization of
glucose and hypoglycemia. Hyperglycemia that is seen
occasionally in the newborn after stress and hypoxia could
result from depression of the glucose regulating mechanisms
through the adrenal medulla.

Qu.De Meyer : What kind of anesthesia do you use in your
animals with sham-operation?
An.Kollee : The sham-operation procedures were performed
under light ether anesthesia.

Qu.Shelley : I was very surprised at the high plasma free
fatty acid concentration in the foetus. May I ask you
how you delivered the foetuses? For instance, in the ligated litters, did you deliver the foetuses from the operated
horn first and then those from the contralateral horn or
were they delivered simultaneously?
An.Kollee : After killing the mother, the foetuses were
delivered within a few minutes. The foetuses from both
ligated and contralateral horns were delivered simultaneously by two persons to prevent prolonged hypoxia in one of
both groups.

DISCUSSION ABOUT THE FIRST SESSION

<u>Qu.Britton</u> : You stated that the administration of insulin
to the sheep foetus increased the umbilical arteriovenous
differences for glucose. Under these conditions does the
glucose uptake exceed that required for oxygen consumption
and growth? If so what is the fate of the glucose?

<u>An.Jones</u> : In our experiments in which insulin has been
administered to fetal sheep we do not have measurements on
the fate of the extra glucose consumed. It would not be
surprising under those circumstances if a substantial proportion of the glucose that is not oxidized was used in
biosynthetic reactions.

<u>Qu.Britton</u> : The growth retarded foetus, produced by res-
triction of uterine blood flow, may be hypoglycaemic but
do we know enough about placental permeability and metabolism of glucose to conclude that the supply of glucose to
the foetus has also been restricted?
When the placenta is perfused <u>in situ</u> through the umbilical
vessels by the recirculating technique, the glucose concentration in the perfusate is lower than that in maternal
plasma. This is presumably due to metabolism of glucose by
the placenta. Changes in either the circulation in the
placenta or in the placental metabolism of glucose may thus
cause the foetal plasma glucose to fall without restriction
of glucose transport to the foetus.

<u>An.Jones</u> : There are no reasonable data on the growth retar-
ded foetus to tell us whether the functional capacity of
this placenta to transfer glucose is less than normal.
The hypoglycaemia that we see in the growth retarded foetal
guinea pig and sheep is associated with a concommitant hypo-
insulinaemia and in these situations the close relationship
between plasma insulin and glucose concentrations is maintained. This suggests that the availability of glucose in
the growth retarded foetus is less than normal. The causes
of this are unknown but are currently being investigated.

Qu. Minkowski : What about brown adipose tissue in the
guinea pig in case of depletion ?
An. Hull : Newborn guinea pigs have both brown and white
adipose tissue. Undersized newborn guinea pigs have a high
or even higher percentage of body fat than their apparently
better nourrished littermates.

Qu. Milner : How long can a pig survive postnatally on
transplacentally derived lipid ?
An. Hull : Not very long, I imagine, perhaps 2 or 3 days.

Com. Hull : Concerning the supply of fatty acid to the
fetus and fat stores in the newborn of different mammalian
species, I would make two comments :
a) Many species can transfer sufficient fatty acid across
   the placenta to meet structural and energy requirements.
   Recently we have found this so in pigs who have very
   little stored fat at birth.

b) In view of Professor Moses' observation, intralipid has
   been infused into pregnant women with the aim of combat-
   ting intrauterine growth retardation. Intralipid does
   supply F.A. which cross the human placenta. The real
   question is, is there evidence that the IUGR infant is
   fat depleted? Retarded infants, even pre-term infants
   at 30 week gestation have sufficient fat in their stores
   to support their energy metabolism for 2 or 3 days.

SESSION II
ENERGY EXPENDITURE

CALORIC NEEDS OF THE NEWBORN
Dr G. Verellen

The outlook for the high risk newborn has improved remarkably over the last 15 years. Advances in the techniques of resuscitation and life support have reduced the mortality from respiratory problems with the result that very low birth weight infants or those suffering from operable malformations of the digestive tract now survive in much increased numbers. Our task is to ensure that these infants are provided with the means not only to survive but to develop physically and mentally to their full genetic potential. Evidence does exist that nutrition plays a major role in achieving this goal.

The objectives of this study have been :
1. to define the minimal amount of calories needed for maintenance in intravenously and orally fed newborns
2. to underline the variable caloric requirements according to the quality of growth in the very low birth weight infant fed by his own mother's milk or by an adapted cow's milk formula.

METHODS AND SUBJECTS
The studies involved direct measurement of
(1) three days intake and output of carbohydrate, protein and fat
(2) the net metabolisable energy intake in Kcal/kg/day as calculated from the measured concentrations of the three principal nutrients in food and excreta
(3) energy expenditure by indirect calorimetry in a thermo-neutral environment (1)

168

(4) urinary nitrogen output
(5) actual respiratory quotient and non protein respiratory
    quotient
From these data, calculations of the oxydation and deposition rates of carbohydrates, protein and fat were made.

The intravenously fed newborns were essentially surgical
cases with a mean gestational age of 36 weeks, birth weight
of 2600 g and post-natal age of 16 days. The patients were
studied after the second post-operative day, when the
physiological functions were stabilized.

Clinical data of the orally fed infants are given in
Table (1)

CLINICAL DATA (M±SD)

	BREAST FED	FORMULA FED
n (studies)	6	13
n (infants)	4	6
GEST.AGE (weeks)	28.8 ±1.7	29.5±1.1
BIRTH WEIGHT (g)	1037±130	1132±127
POSTNATAL AGE (weeks)	1-4	1-4
MEAN CALORIE INTAKE (Kcal Kg day)	93.4± 24	127.7 ±26

Table 1

RESULTS
The relationship between endogenous fat oxydation and
caloric intake in newborns receiving a fat free intravenous
solution of Glucose or Glucose-amino acids mixture is
given in fig. (1). A highly significant inverse relationship exists, indicating that up to 70-80 Kcal/kg/day intake,
the newborn does oxidize his own fat. This implies that
the minimal caloric requirement for maintenance is around
70 Kcal/kg/day.

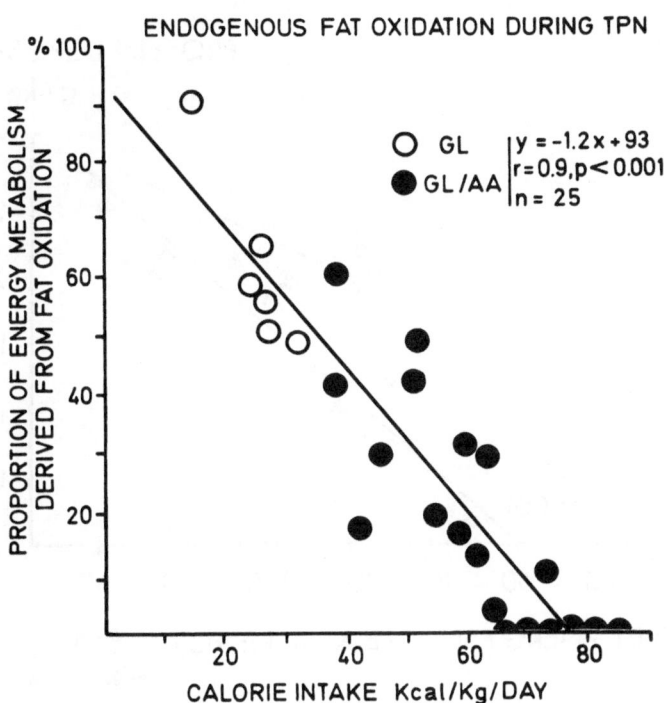

Fig. 1. Correlation between net caloric intake and
proportion of energy metabolism derived from
fat oxydation in neonates fed intravenously
with glucose solution (O) or glucose-amino
acid solution (●).

In the formula and breast milk fed infants, the relationship between protein and fat deposition, and net energy
intake were examined by means of linear regression.
Nitrogen retention correlates positively with net energy
intake (fig. 2) as expected. Not so expected was the
finding that breast milk fed infants deposited twice as
much nitrogen than the formula fed infants at the same
level of net energy intake.

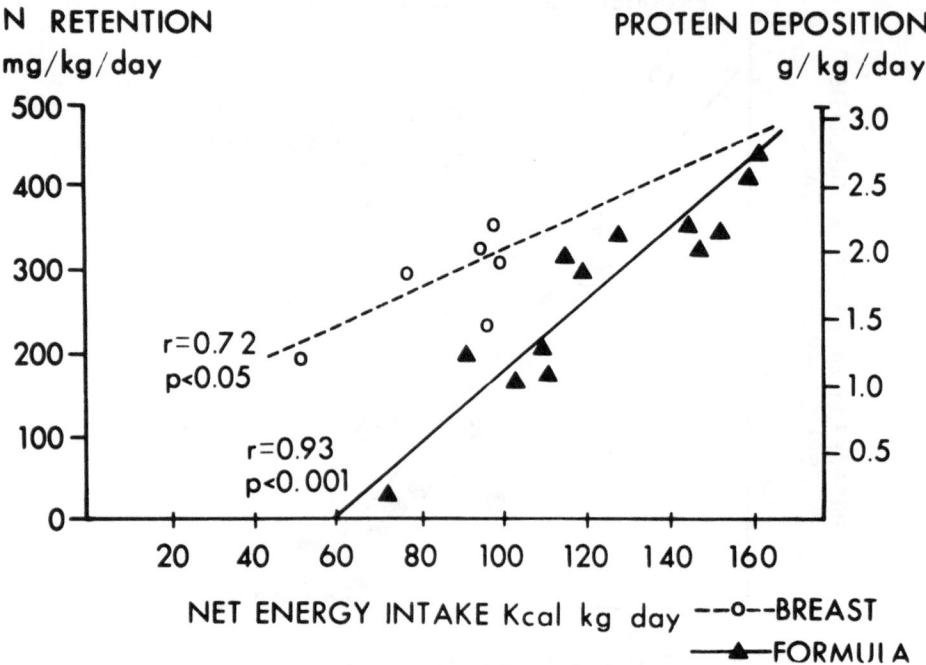

Fig. 2.

Correlation between net energy intake and nitrogen retention
in breast fed (O---O) n=12;r=0.70; p< 0.015) and formula fed
(▲——▲) n=13;r=0.88; p< 0.001) low birth weight infants.

Fat deposition increases with increasing net energy intake
in both groups (fig. 3), but at the same caloric intake
more fat is retained by formula fed infants as opposed
to those fed by their own mother's breast milk. The net fat

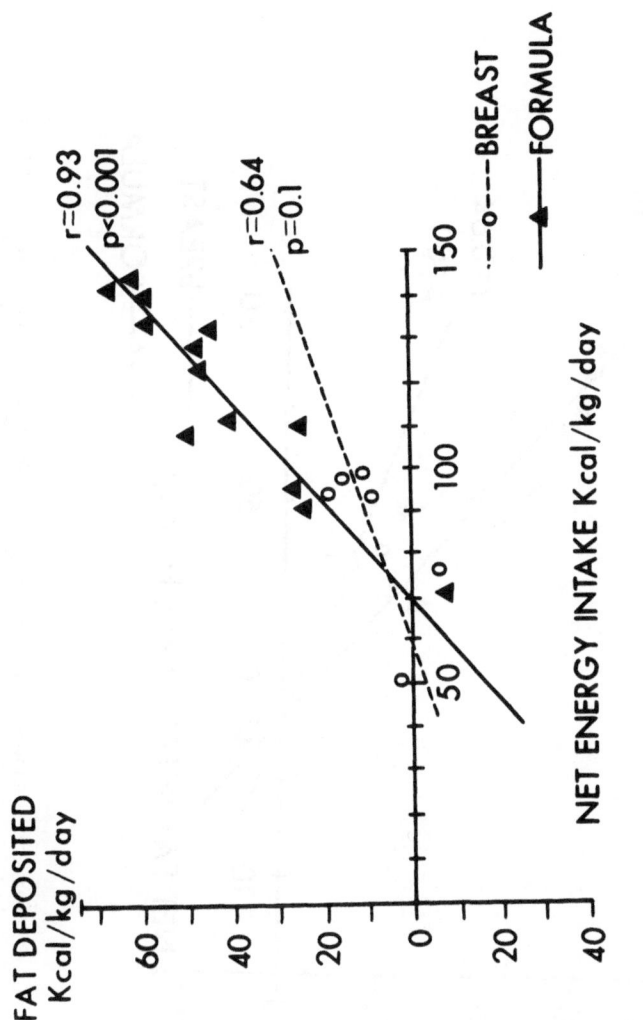

Fig. 3. Correlation between net energy intake and tissue deposition
of fat in breast fed (0--─-0 n=12; r=0.54; p < 0.05) and formula fed (▲──▲ n=13;r=0.92; p < 0.001) low birth weight infants.

intake and fat deposition demonstrate the same relationship indicating that better fat absorption does not account
for the increased fat retention observed in the formula
fed group (fig. 4).

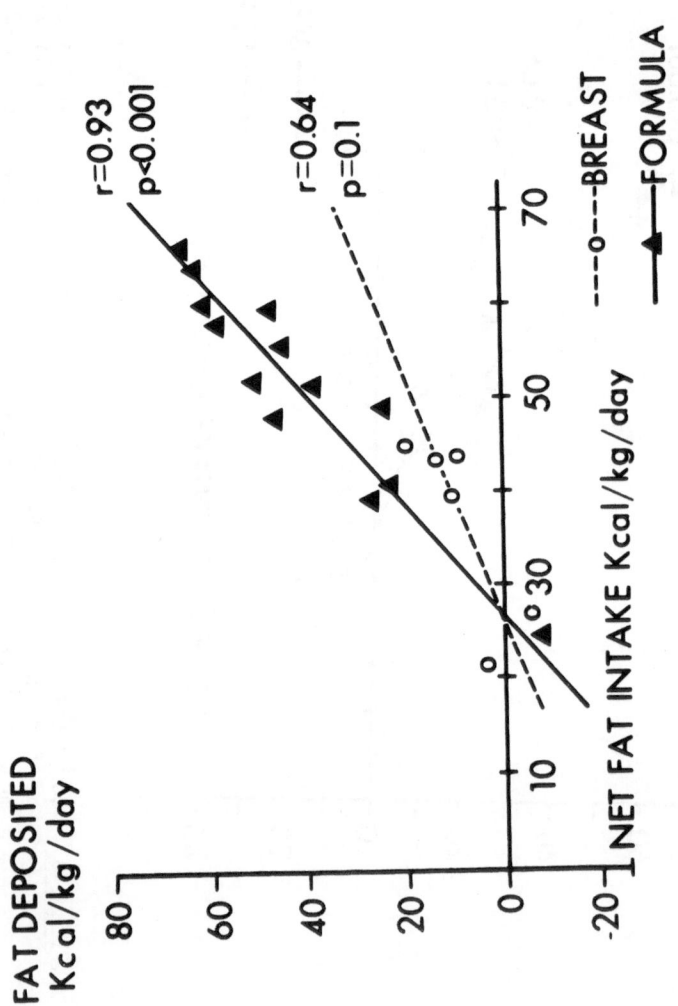

Fig. 4. Correlation between net energy intake from fat and tissue deposition of fat in breast fed (o--o); n=12; r=0.70; p<0.01) and formula fed (▲—▲; n=13;r=0.93; p<0.001) low birth weight infants.

Once the net energy intake is less than 70 Kcal/kg/day
(fig.3), the infants oxidize their own lipid stores.

DISCUSSION

Both in intravenously and orally fed newborns, a net energy
intake of less than 70 Kcal/kg/day results in endogenous
fat oxydation. This suggests that it is the minimal amount
of calories required to cover basal energy expenditure and
muscular activity in a thermoneutral environment. This
figure is in good agreement with estimates of caloric expenditure provided by Sinclair (2).

The comparison between the infants fed by their own mother's
milk or formula suggests that the quality of growth
(protein versus fat deposition) is different : at the same
net energy intake, more protein and less fat is deposited
by the breast milk fed infant. The intra-uterine nitrogen
retention rate of approximately 325 mg/kg/day can be obtained by 100 Kcal/kg/day net energy intake derived from the
infant's own mother's milk and by 145 Kcal/kg/day net
energy intake derived from formula.
These results illustrate the changing caloric requirements
for growth according to the specific diets offered.
The studies also support the recent results of Atkinson
et al (3-6) demonstrating that very low birth weight infants might benefit from being fed by their own mother's
early milk.

REFERENCES

1. Swyer, P, G Putet, JM Smith, T Heim, Energy metabolism and substrate utilization during total parenteral nutrition in the Newborn. In : Intensive care in the newborn. II. S. Stern, W. Oh, B Friis-Hansen, Eds. Masson Publi - shing U.S.A. Inc. (N.Y.) 1978, pp 307-316.

2. Sinclair, JC, Energy balance of the newborn. In : Temperature regulation and energy balance in the newborn. Sinclair, JC, Ed. New York, Grune and Stratton, 1978, pp. 187-203.

3. Atkinson, SA, MH Bryan and GH Anderson, Human milk : Difference in nitrogen concentration in milk from mothers of term and premature infants. J Pediatr 93: 67-69, 1978.

4. Atkinson, SA, GH Anderson and MH Bryan, Human milk : Comparison of the nitrogen composition in milk from mothers of premature and full-term infants. Am J Clin Nutr 33:811-815, 1980.

5. Atkinson, SA, IC Radde, GW Chance, MH Bryan and GH Anderson, Macro-mineral content of milk obtained during early lactation from mothers of premature infants. Early Hum Dev 4:5-14, 1980.

6. Atkinson, SA, MH Bryan and GH Anderson, Energy content of human milk during early lactation from mothers giving birth prematurely and at term. Am J Clin Nutr (In press, 1980).

LIST AND TITLE OF THE CONTRIBUTORS TO THIS WORK

1. G. Putet and G. Verellen, M.D., Research Fellows,
   Neonatal Division, Hospital for Sick Children, Depart-
   ment of Pediatrics, Medical Faculty, University of
   Toronto, Toronto Ontario, Canada.

2. T. Heim, M.D., Ph. D., Visiting Professor of Pediatrics
   and Developmental Biology, Research Institute of Hospital for Sick Children, Department of Pediatrics, Medical
   Faculty, University of Toronto, Toronto Ontario, Canada.

3. J.M. Smith, Ph. D., M.A. Sc., Director of the Department
   of Medical Engineering, Hospital for Sick Children,
   University of Toronto, Toronto Ontario, Canada.

4. P.R. Swyer, M.B., F.R.C.P. (Lon.), Professor of Pedia-
   trics, Director, Neonatal Division, Hospital for Sick
   Children, Department of Pediatrics, Medical Faculty,
   University of Toronto, Toronto Ontario, Canada.

5. S.   Atkinson, Ph. D., Department of Nutrition and Food
   Science, Medical Faculty, University of Toronto, Toronto
   Ontario, Canada.

6. H.G. Anderson, Ph. D., Professor of Nutrition, Department of Nutrition and Food Science, Medical Faculty,
   University of Toronto, Toronto Ontario, Canada.

ACKNOWLEDGMENTS

The project was supported by grants from National Health
and Welfare of Canada No.606-1482 and Physicians Incorporated Foundation No.9859. Dr. Gaston Verellen was holding a
NATO Fellowship and Stephanie Atkinson a Hospital for Sick
Children Foundation Studentship during the studies.

DISCUSSION

Qu. Milner : What was the formula ?
An. Verellen : Sma 20 in the first week of life, and Sma
24 afterwards.

Qu. Milner : What were the fecal lipid losses in breast
feeding versus formula ?
An. Verellen : Breast milk and formula fed infants received
respectively 5,3 $\pm$ 1,38 and 5,61 $\pm$ 0,6 g/kg/day of lipids
in the diet. The losses were, respectively, 19,4 % and
14 % of the intakes.

Qu. Milner : Should be reserved about results from breast
fed since n=6 and 4/6 are grouped close to 100 kcal/kg/day?
An. Verellen : I fully agree that these results are preliminary and should be confirmed in the future, especially
for fat deposition at high net energy intakes in the breast
milk fed infants.

Qu. Siegrist : Could you comment further on the composition of the formula milk used? I am asking you because
breast milk contains a considerable amount of cholesterol
and adapted cow's milk normally does not. Could the differences in retention and burning between breast and formula
fed babies be due to this difference ?
An. Verellen : That could be the explanation. Another possibility is that the higher nitrogen retention observed in
the breast fed babies requires extra calories and that
this energy cost of growth is provided by fat oxydation.

Qu. Moore : R.Q. measurement is subject to error and the
fat deposition is calculated from the R.Q. What were the
spread of values in any study ?
An. Verellen : R.Q. value varies considerably with the
type of fuel oxidized but in a steady metabolic state,

175

repeated measurements of the respiratory quotient during
quiet sleep show a spread of less than 5 %.

Qu. Eggermont : Did you measure skinfold thickness in your
babies fed either human milk or formula ?
An. Verellen_ : We did measure skinfold thickness in these babies but, unfortunately, the results have not been
analysed at the present time.

VARIATIONS IN THE RATE OF OXYGEN CONSUMPTION OF NEWBORN HUMAN INFANTS

Prof David Hull

The variations in the rate of oxygen consumption of newborn infants have been reported in relation to body weight (1), gestational age (2), post-natal age (3), ambient temperature (4), activity (5), nutrition (6) and sleep state (7). The information has been used by clinicians to help determine the nutrient requirements and the appropriate ambient temperatures to provide thermal comfort. Although the data has been of value as a general guide, its use in the preparation of management programs for individual infants has been limited, for, even after making allowances for all the known factors there is still considerable variation in the rates of oxygen consumption and thus the nutrient energy requirements and rates of heat production of apparently similar infants. This is true for the mature healthy infant and also for the sick preterm infant who might be thought to benefit most by close attention to food energy intake and operative ambient temperatures. In this paper I will briefly review those factors which influence metabolic rate and the magnitude of their effect, then I shall review the variance observed in apparently similar infants under usual nursing conditions, and discuss the implications of this variation, and finally speculate of the possible mechanism involved.

THE MAGNITUDE OF CHANGES PRODUCED BY INFLUENCES
KNOWN TO AFFECT THE RATE OF OXYGEN CONSUMPTION

The rate of oxygen consumption in the human infant has been measured by a variety of techniques over the years, in many of the   earlier studies the possible effects of feeding, the sleep state and even the ambient temperature were not always appreciated;   thus because of methological and technical differences and the varying states of the subjects, comparisons between the absolute values   obtained   by different research workers may not be justifiable, nevertheless there is surprising agreement on resting or minimal values.

The resting rate is on average about 5.5 ml $O_2kg^{-1}min^{-1}$ on the first day rising over the first week or so to just over 8.0 ml O_2kg^{-1} min^{-1}. (8)(9) This rise in metabolic rate occurs in mature and pre-term infants alike, but appears earlier in the hungry 'light for dates' infant and is more gradual in the reluctant to feed pre-term   infant.

Many investigators have studied the effect of feeding on the rate of oxygen consumption. (10)(11)(6) Most have reported a rise varying from mean values of 10 - 15% to as much as 30 - 50%. Many report it to be an extremely variable phenomenon, not occurring in some infants and being marked in others for no obvious reason. Observers also differ as to the timing of the rise, some reporting a peak about 30 minutes after the feed, others after an hour; the phenomenon may well   very with feeding practice and the infant's plane of nutrition. It   has been observed after feeds of human breast milk and a variety of artif-icial cows milk feeds, and individual oral and intravenous nutrients.

The rate of oxygen consumption also varies with changes in sleep state. This too occurs in preterm (32 - 36 weeks gestation) and term infants alike. In a recent study Stothers and Warner (1979) (6),found on average an increase of 15% in preterm infants when they   changed from 'non-rapid-eye-movement' sleep.

Assessment of the possible effects of muscular work   on   the metabolic rate of newborn infants is fraught with problems. The fact that newborn infants spend much of their day sleeping, and in   deep sleep at that, has permitted many of the investigations quoted in this paper. Preterm infants spend even more of their day asleep,   Mestyan et al (1968) (12) found that preterm infants when they were 2, 2 - 15 and 16 - 30 days old respectively, spend on average 84, 75   and   65 percent of their time asleep. Muscular activity over a 24 hour period in infants swaddled and in a thermal neutral environment probably uses less than 10% of the oxygen consumed. But with periods of   vigorous

movement with crying, Rubecz and Mestyan (1973) (5) found that preterm infants, on average could increase their rate 50% (about 3 ml
O_2 $kg^{-1}min^{-1}$) in the first days of life and by 75% (about 6ml O_2 kg^{-1}
min^{-1}) when they were over two weeks of age. Here again, as might be
expected, there was considerable individual variation.

The principles involved in the relationship between the operative
ambient temperature and the infants rate of heat production, which in
states of thermal equilibrium relates directly to the infants rate of
oxygen consumption are illustrated diagrammatically in the figure 1.
The thermoneutral zone is defined as that range of ambient temperature
within which metabolic rate is at a minimum and within which temperature regulation is achieved by non-evaporative physical processes

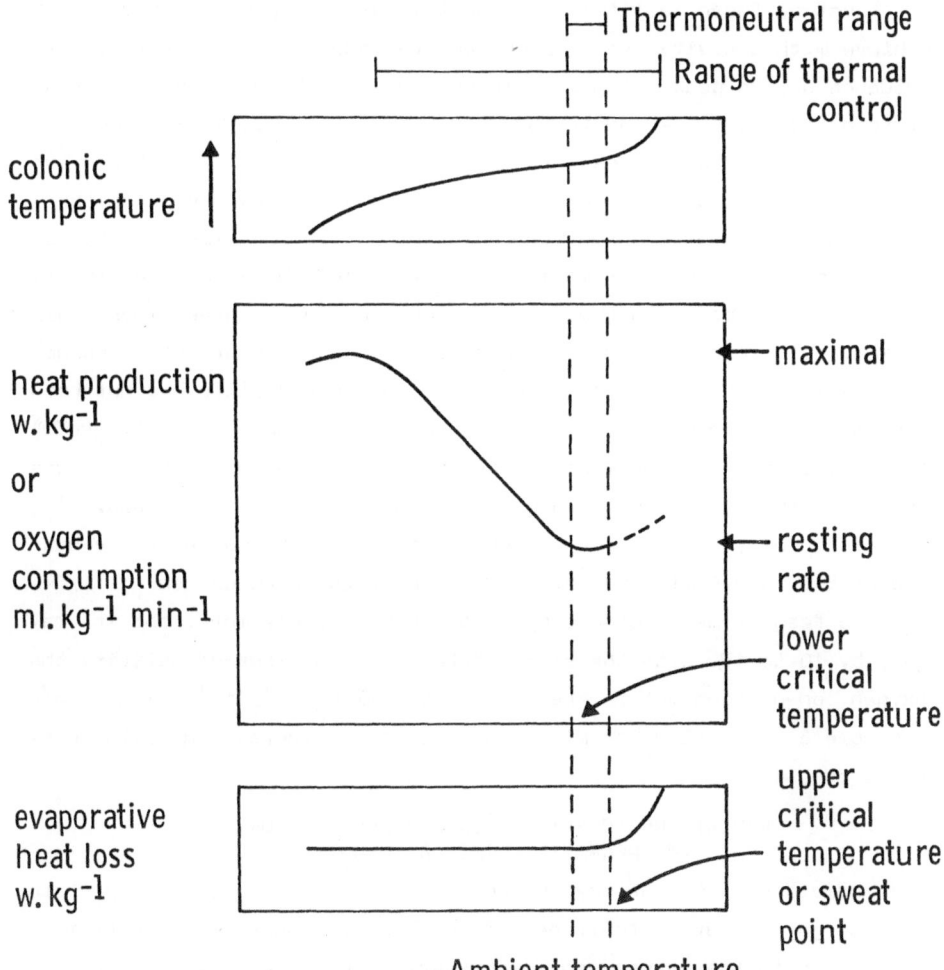

Figure 1: Diagrammatic representation of the relationship
between colonic temperature, metabolic rate and evaporative
heat loss and ambient temperature.

alone. These processes include alterations in peripheral   vasomotor
tone and changes in posture which vary the area of the exposed cooling
surfaces. Such factors operate in the mature newborn infant,   but
their importance in preterm is difficult to assess. The glossary   of
terms for thermal physiology (13) defines minimal observed   metabolic
rate as that measured during specified periods of minimal activity,and
resting metabolic rate as that measured whilst the subject is   resting
in a thermoneutral environment but not in the post—absorptive   state.
Most of the values in the literature related to newborn   infants   are
'resting' or 'minimal' rather than basal metabolic rates. It is often
difficult to determine with certainty that an infant is in a   neutro-
thermal environment.

     Just as there is little data on the maximum rates   babies   can
achieve with activity, so, for obvious reasons,   there is little
evidence on the peak rates on cold exposure. The majority of infants
preterm (32 – 36 weeks gestation) and term infants alike, respond   to
cold exposure by an increase in metabolic rate. The   magnitude   of
individual responses to 'cool' temperatures vary. What evidence there
is suggests that on average mature infants after the first day or   so
can at least double their metabolic rates   (14). Many newborn mammals
can achieve rates of thermogenesis which are two or three times their
minimal rates within minutes of birth. The development of   thermogenesis in human infants over the first 24 hours of life has not been
systematically investigated. In mature infants Smales and Kime (1978)
(15) found that the thermogenic response to a mild cold stimulus   was
initially small and increased gradually over the first 24 hours.

     In summary, the rate of oxygen consumption increases from a mean
around 5.5 to 8.0 ml $O_2kg^{-1}min^{-1}$ over the first week or   so of life.
After a feed it may increase by 0 to 30% but rarely more, and it   may
vary by 10 to 15% with the sleep state. With periods of activity the
oxygen consumption may increase by 3 to 5 ml $O_2kg^{-1}min^{-1}$, with   cold
exposure 6   to 8 ml $O_2kg^{-1}min^{-1}$ both responses increasing   with postnatal age.

VARIATIONS IN RATE OF OXYGEN CONSUMPTION
IN APPARENTLY SIMILAR INFANTS

The individual rates of oxygen consumption of sleeping,   preterm
infants just prior to their next feed in known ambient   temperature
conditions in the nursery show wide variations (Figure 2) (16). This
variation is common to most studies, for example Rubecz and   Mestyan

(1975) (5) again in preterm infants show ranges 4.0 to 9.0 (mean 5.8), 4.0 to 10.0 (mean 6.8) and 4.8 to 15.0 (mean 7.9) ml O_2 $kg^{-1}min^{-1}$ in infants under 2 days, 2 – 15 days and 16 – 28 days respectively. Thus the minimal rate of one infant may be as great as that of another during intense activity or maximally responding to cold exposure!

Serial measurements over 24 hours, and the first three weeks of life show that in general those infants with high rates maintained high rates, and those infants with low rates maintained low rates (Figure 3). (9) There was no apparent difference in their growth rates but there was in their milk intake.

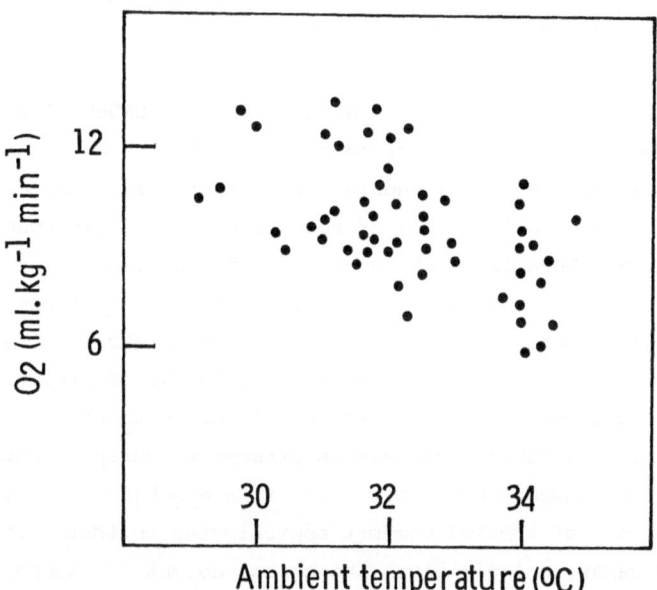

Figure 2: Variation in metabolic rate at ambient temperatures close to the lower critical temperature in healthy preterm newborn infants of similar weight and postnatal age. The measurements were made in the nursery, whilst the infants were asleep about 2 hours after a feed using an open circuit method(19)

184

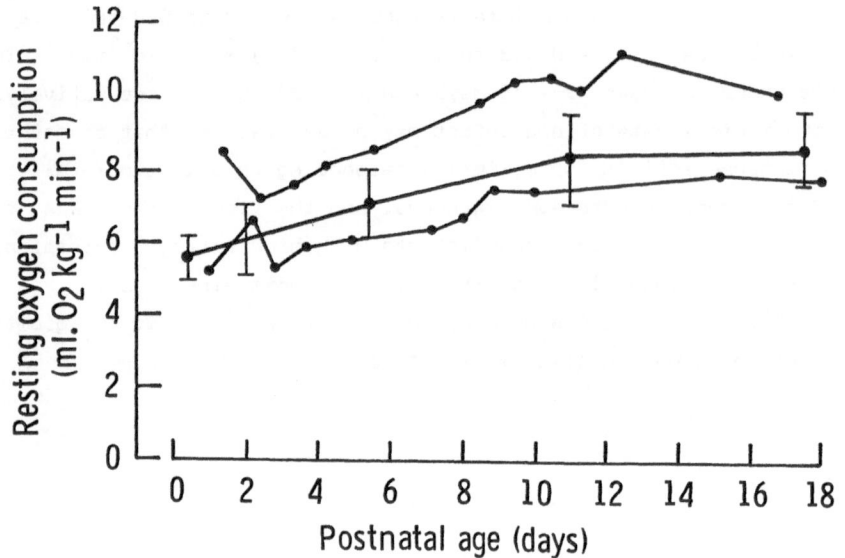

Figure 3: The rise in metabolic rate after birth of healthy
preterm infants. Two individual infants are illustrated, one
with a low, the other with a high rate (9).

IMPLICATIONS OF VARIATIONS IN RATES OF OXYGEN CONSUMPTION.

The infants energy requirement for maintenance will be met when his
intake of metabolizable energy equals his rate of heat   production.
When his food intake is in excess of requirements for maintenance and
thermoregulation, there is a net increase in the bodies energy content
in the form of protein and fat. In other words the infant begins to
grow. The significance of the wide variation in metabolic   rates of
infants is that some will need twice the food intake of others to meet
maintenance requirements and before they will begin to grow.

    It is a reasonable clinical aim to attempt to nurse infants   in
an environment of thermal comfort. Comfort is a subjective   matter;
in adults the zone of thermal comfort approximates to that   of   the
thermoneutral range, animals given the choice select   temperatures
around, sometimes just below, the lower point of the range.   In   the
newborn with their capacity to limited sweat,   hot temperatures are
to be avoided; likewise by virtue of their relative large   surface
area and poor thermal insulation a small fall in ambient temperature
below the thermoneutral range may make unacceptable demands for extra
thermogenesis to maintain thermal stability.   The   unpredictable
variation in minimal metabolic rate makes keeping all   infants   in
thermal comfort an exacting task. The features that determine   the

thermal balance between a naked infant and his environment are such that, at the ambient temperature that would suit those infants with metabolic rates close to the mean, occasional infants will experience considerable cold where as others would be stimulated to sweat. (Figure 4).

SPECULATIONS ON FACTORS CONTRIBUTING TO INDIVIDUAL VARIATIONS IN THE RATE OF OXYGEN CONSUMPTION

It would be inappropriate here to attempt to review the extensive literature on the factors which influence the basal metabolic rate and energy requirements for maintenance in adult man and animals, suffice it to say that much of the variance remains unexplained. It may well reflect genetically determined differences such as the bodies make-up or variation in energy efficiency at a cellular level. That genetically determined factors are important is supported by the observation that differences are present as soon as they can be measured after birth. However there are other possibilities. For example the variation might be due to differences in the thermal exchanges with the environment. Thus although the infants were of a similar age

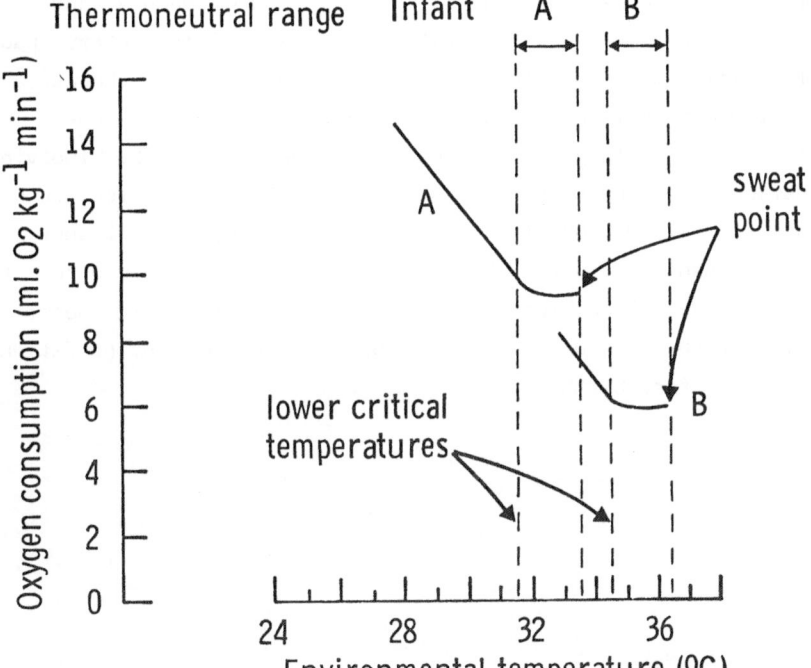

Figure 4: Diagrammatic illustration of the difference in the thermoneutral ranges between infants with high and low minimal metabolic rates which might be expected from our current understanding of their thermoregulatory responses.

maturity, and body weight, they may differ in the state of their peripheral vasomotor control or their usual posture and this would influence the rates of heat loss by radiation, convection and conduction. Those with higher rates of heat loss would be stimulated to produce extra heat and thus have higher rates of oxygen consumption in identical ambient conditions. From what we know of the changes in vasomotor control and the effects of posture, even if these two mechanisms are responsible for the width of the thermoneutral range, it is unlikely that differences between one infant and another could explain the observed variations in oxygen consumptions. However evaporative heat exchange due to transepidermal water loss may be very large in the first days of life in mature infants and for many days in the preterm infant (17). In the very immature evaporative heat loss may exceed the minimal rate of heat production for some days after birth. Although this might explain some of the variation in studies on infants in the first days of life, it cannot explain its persistence beyond the second week.

Two environmental factors have been shown to increase the minimal metabolic rate of adults, prolonged cold exposure and over feeding. Recent evidence from rats has shown that in dietary induced thermogenesis as in cold induced thermogenesis, much of the extra heat production occurs in brown adipose tissue. (18) It has been suggested that dietary induced thermogenesis protects the individual against obesity, and that those who become obese have defective brown adipose tissue. Newborn infants are not cold adapted, but they do have thermogenically active brown adipose tissue. It could be that some of the variation might be due to factors which influence the activity of this tissue including possibly, the level of nutrition of the infant before birth.

REFERENCES

1. Sinclair, JC, Metabolic rate and body size of the newborn. Clin. Obstet. Gynecol. 14:840-854, 1971

2. Bruck, K, Temperature regulation in the newborn infant.   Biol. Neonate 3:65-119, 1961

3. Hill, JR and DC Robinson, Oxygen consumption in normally   grown small-for-dates, and large-for-dates new-born infants.   J of Physiol 199:685-703, 1968

4. Hey, EN, The relation between environmental temperature and oxygen consumption in the new-born baby. J. of Physiol 200: 589-603, 1969

5. Rubecz, I and J Mёstýan, Activity, energy metabolism and   postnatal age relationship in low-birth-weight infants.   Acta Paediatr Acad. Sci Hung. 16:351-362, 1975

6. Stothers, JK and RM Warner, Effect of feeding on neonatal oxygen consumption. Arch Dis Childh 54:415-420, 1979

7. Stothers, JK and RM Warner, Oxygen consumption and neonatal sleep states. J of Physiol 278:435-440, 1978

8. Bhakoo, ON and JW Scopes, Minimal rates of oxygen consumption in in small-for-dates babies during the first week of life. Arch Dis Childh 53:850-854, 1978

9. Rutter, N, SM Brown and D Hull, Variations in the resting oxygen consumption of small babies, Arch Dis Childh 53:850-854, 1978.

10. Mёstýan, J, I Járai, M Keketc and   Soltesz, Specific dynamic action in premature infants kept at and below the neutral temperature, Paediat Res 3:41-50, 1969

11. Gentz, J, M Kellum and B Persson, The effect of feeding on oxygen consumption, RQ and plasma levels of glucose, FFA and D-B hydroxybutyrate in newborn infants of diabetic mothers   and small for gestational age infants, Acta Pediat Scand 65:445-454, 1976

12. Mёstýan, J, I Járai, M Feketc, The total energy expenditure and its components in premature infants maintained under different nursing and environmental conditions, Pediatr Res 2:161-171, 1968

13. Bligh, J, and KG Johnson, Glossary of terms for thermal physiology, J of Appl Physiol 35:941-961, 1973

14. Hey, EN, Physiological control over body temperature in: Heat loss from Animals and Man, Monteith JL, LE Mount (Eds) London

14. Hey, EN, Physiological control over body temperature in: <u>Heat Loss from Animals and Man</u>. Monteith JL, LE Mount (Eds) Butterworth 1974

15. Smales, ORC, and R Kime, Thermoregulation in babies immediately after birth, Arch Dis Childh 53:58-62, 1978

16. Smales, ORC, A simple method for measuring oxygen consumption in babies, Arch Dis Childh 53:53-58, 1978

17. Rutter, N and D Hull, Water loss from the skin of term and preterm babies, Arch Dis Childh 54:858-868, 1979

18. Rothwell, WJ and MJ Stock, A role for brown adipose tissue in diet induced thermogenesis, Nature 281: 31-35, 1979

19. Smales, ORC, A simple method for measuring the rate of oxygen consumption in the nursery, Arch Dis Childh 269:53-57, 1978

RECENT GENERAL REFERENCES

Mount, LE, Adaptation to Thermal Environment. Edward Arnold 1979

Sinclair, JC, Temperature regulation and energy metabolism in the newborn, Grune and Stratton 1978

DISCUSSION

Qu.Minkowski : What about O_2 consumption and heat loss in light for dates infants?

An.Hull : The minimal rate of oxygen consumption of light for dates expressed per kg. body weight is similar to that of other infants in the first hours after birth but increases more rapidly and, say, higher over the first week of life than infants of similar birth weight.

Bhakoo,O.N. and Scopes, J.W. (1974). Minimal rates of oxygen consumption in small-for-dates babies during the first week of life. Arch.Dis.Child. 49, 583-584.

Rutter, N., Brown, S.M. and Hull, D. (1978). Variations in the resting oxygen consumption of small babies. Arch.Dis. Child. 53, 850-854.

Qu.Teller : Do premature babies with high physical activity show increased concentrations of thyroid hormones in the blood? Also would low body temperature tend to increase thyroid hormone levels and thyroid activity?

An.Hull : I do not know of any studies on preterm infants in which physical activity or low body temperatures have been related to thyroid function.

Qu.Teller : Is the transcutaneous water loss correlated to thickness of skin. In other words : do more immature neonates show a higher degree of water loss than more mature ones?

An.Hull : Transepidermal water loss is related to maturity, and is very high in infants under 28 weeks gestation. This variation is probably related more to the permeability of the skin cell layers to water than to the thickness of the skin.

Qu.Moore : I was delighted to see Professor Hull emphasizing the difference between individual and the group mean. This has been recognised since the work of Herington in 1942, but has never been adequately analyzed.

The potentially high rate of water loss in the premature

babies is surprising. It is comparable with maximal swea-
ting rate in the adult. Would Professor Hull indicate what
methods he used to restrict the transpired water loss?
An.Hull : Three methods have been used to reduce water loss
in the clinical situation. The first is to cover infants
with a plastic blanket. This, if it is applied to the
babies' skin stops evaporation and is probably undesirable.
However a 'bubble' blanket around the baby creates a micro-
environment with high humidity. The second is to raise the
water content in the ambient air. The third is to reduce
water loss through the skin by applying appropriate creams.
Over the centuries newborn babies have been massaged with a
wide variety of oils and ointments. We have just completed
a control trial comparing the effects of an ointment and a
plastic 'bubble' blanket on transepidermal water loss from
premature infants.

Qu.De Meyer : Can the water loss in animals be prevented by
a 100 % humidity environment?
An.Hull : Transepidermal water loss will be reduced to a
minimum if the animal is in ambient air fully saturated at
a thermoneutral temperature.

Qu.Colombo : It is amazing how large the transepidermal water loss is in the premature infant. One would assume that
also other constituents may leak out also. This may influence the treatment and feeding of these babies. What information do you have on this point?
An.Hull : We have no information on what other constituents
leak out with the water. It is a very interesting question.

ENERGY REQUIREMENTS FOR GROWTH IN THE NEONATE

P.J.J. Sauer, H.J. Dane, R.G. Pearse and H.K.A. Visser

Introduction

Although a great deal is known about the energy cost of
growth in animals and in infants recovering from malnutrition, very little is known about it in neonates. The main
reason for this seems to be that this type of research in
neonates is very difficult.

Energy cost of growth can be divided into two parts, the
energy present in the components of new tissue and the
energy necessary to organize these components into the
new tissue:

E(cost of growth) = E(components) + E(synthesis).

The energy present in the components will be stored in
new tissue, together with part of the energy necessary to
organize the new tissue.

The energy cost of growth can be estimated in the following
ways:

1. by using the data from carcass analysis
2. by calculating the energy balance
3. from the increase of oxygen consumption after a feed
   (specific dynamic action)
4. from the number of A.T.P. equivalents necessary for
   growth
5. from studies in the animal model

With carcass analysis only the energy present in the components of new tissue can be calculated. In combination

with data for E(synthesis) derived from animal models it
is possible to calculate the total cost of growth. The
energy present in the components of new tissue can be calculated from the energy balance, the cost of tissue synthesis has been calculated indirectly from the energy
balance. With methods 3 and 4 the cost of tissue synthesis
can be calculated.
Theoretically it is possible to calculate both parts of the
energy cost of growth from animal models.

Carcass analysis
Widdowson (1) and Ziegler (2) performed extensive analysis
of neonates who died just before or directly after birth.
From these figures Ziegler has calculated the so called
"reference fetus", providing the average body composition
of a fetus of a given weight and gestational age. It is
possible to calculate the energy cost of components of
growth from the increase in fat and protein content of this
"reference fetus". In combination with data from animal
studies it is possible to calculate the total cost of
growth. A criticism of this method is, that the "reference
fetus" is calculated from measurements done in infants
who had died before or directly after birth. No postnatal
growth had taken place and it is not known if and how much
of the energy stores are depleted just before death.

Energy balance
The energy balance equation has been used by different
authors to calculate the energy cost of growth, both in the
neonate (Sinclair (3), Sauer et al. (4), Brooke (5)), as
well as in older infants recovering from malnutrition
(Ashworth (6), Spady et al. (7)). The energy balance
equation for a growing neonate can be written as follows:
E(intake) = E(excreted) + E(expended) + E(components).
E(intake) represents the energy in the food.
E(excreted) represents losses via faeces, urine and vomiting.
E(expended) comprises 1. the energy used in maintenance,
2. the energy used for thermoregulation, 3. the energy used

in activity, 4. the energy used for synthesis of new tissue
when E(expended) is measured by indirect calorimetry and
a part of this energy when it is measured by direct calorimetry.
E(components) is the energy inherent in the components
of new tissue.
The energy of work is omitted from the equation, because
although neonates move frequently all the energy will be
dissipated as heat.
The optimal energy intake for a neonate is not well known,
but would seem to be in the range 500 to 600 kJ/kg/24 hr.
The energy present in the food can be calculated from the
constituents or can be measured using a bomb calorimeter.
The losses via faeces, urine and vomiting can also be calculated from the constituents or be measured directly.
Energy losses via the urine are small. It is known from
various investigations that faecal losses depend on the
composition of the food, particularly on the fat composition
Medium chain triglycerides and polyunsaturated fatty acids
are better absorbed by the neonate than long chain saturated fatty acids. Normally 10-20% of the intake will be lost
via the faeces.
E(expended) comprises the energy used in maintenance,
activity, thermoregulation and tissue synthesis. The
energy required for maintenance is the energy necessary
for maintaining the body without a change in body composition. Some authors have called this basal expenditure. In
neonates it is not possible to measure basal energy expenditure following the criteria used for adults, because it
requires withholding feeding for 18 hours. Thus measurement
of basal energy expenditure is not suitable for neonates.
Minimal observed metabolic rate in growing infants will be
higher than maintenance energy expenditure, as the cost of
tissue synthesis will be included in the minimal observed
metabolic rate. E(expended) has been measured by both
direct and indirect calorimetry. Although both methods are
physically totally different, it has been assumed that they
are physiologically identical since the studies of Howland

(8) in 1913 and Day (9, 10) in 1942 and 1943. Direct calorimetry measures the amount of heat dissipated by the body via convection, conduction, radiation and evaporation. An increase or decrease in body temperature must also be taken into account.

In indirect calorimetry the amount and composition of foodstuffs combusted are calculated from the amount of oxygen consumed and carbon dioxide produced. An alternative direct method - the determination of insensible weight loss - is not a reliable approach to the determination of total energy expenditure since a standard proportionality between evaporative and non-evaporative heat losses cannot be assumed in neonates.

When all food that is combusted is given off as heat, the outcome of indirect calorimetry will be equal to that of direct calorimetry. However, when a part of the energy released in the body is not given off as heat but stored in the body without increasing the body temperature, the outcome of indirect calorimetry will be higher than that of direct calorimetry.

Recently Brooke et al. (5) published a study on the energy-cost of growth in low birth weight infants, using the energy balance. E(expended) was measured by indirect calorimetry during several periods of different activities. E(components), calculated using the energy balance, was 16.8 kJ/g growth. Using the regression analysis of weight gain against energy retention total cost of growth was calculated as 24 kJ/g growth. This means that the cost of tissue synthesis is 7.2 kJ/g growth. Using the same methods Spady et al.(7) calculated the energy cost of growth in infants recovering from malnutrition. They found the total energy cost of growth to be 18.4 kJ/g growth, divided into 13.8 kJ for components and 4.6 kJ for synthesis.

Specific dynamic action

Brooke and Ashworth (11) showed an increase in oxygen consumption after a feed in infants recovering from malnutrition. The more the postprandial increase was greater, the

more rapid the rate of growth. It was assumed that this
increase in oxygen consumption represented the cost of
growth, being the cost of tissue synthesis, as the food
was oxidized. Using this method the cost of tissue synthesis can be calculated as 1.1 kJ/g growth.

A.T.P. equivalents

Using Atkinson's metabolic price system, Hommes et al. (12)
calculated the energy cost of growth. They used data from
the literature on changes in body composition. From the
amount of A.T.P. equivalents necessary for growth, they
found E(components) to be 6.8 kJ/g growth and E(synthesis)
1.3 kJ/g growth.

Table 1. Energy cost of growth (kJ/g growth)

	components	synthesis	total
Widdowson (1), Kielanowski (15)	6.0	2.5	8.5
Ziegler et al. (2) Kielanowski (15)	6.4	3.0	9.4
(body composition data (1), (2); data from animal models (15))			
Sauer et al. (4)	11.9	1.1	13
Brooke et al. (5)	16.8	7.2	24
Spady et al. (7)	13.8	4.6	18.4
(data from energy balance studies)			
Brooke and Ashworth(11) (measuring specific dynamic action)		1.1	
Hommes et al. (12) (calculating "metabo- lic price system")	6.8	1.3	8.1

Our own studies

In six infants weighing less than 2.5 kg the energy cost of
growth was calculated using the energy balance.

E(expenditure) was measured with direct and indirect calorimetry. Some of these results have been published (Sauer
et al. (4)).

Using a new calorimeter in which it is possible to do
direct and indirect calorimetry continuously and simultaneously for several hours, E(expended) was measured several
times in babies weighing less than 2,5 kg during a period
when they were growing rapidly. Birth weight varied between
920 and 1400 g and gestational age at birth from 29 to 33
weeks. The babies weight at the time of the study varied
between 870 and 2100 g and the age from 8 to 58 days. The
patients were growing rapidly during the period in which
the measurements were made; their growth curve for weight
was below the -2 SD growth curve of Usher and McLean (13),
but parallel to this curve. These babies were therefore
all small for gestational age at the time of study.

34 studies were made; the periods of measurement varied
from 4 to 24 hr. in all cases and were 6 hr. in 28 cases.
During the weeks in which these studies were done and
during the measurements themselves the babies received
Nenatal as a continuous infusion into the stomach via a
nasogastric tube, ± 175 ml/kg/24 hr. The composition of
Nenatal is given in table 2.

Table 2. Composition of Nenatal (Nutricia) per 100 ml
formula

Milk fat	0.1 g
Vegetable fat	4.4 g
(Linoleic acid)	1.4 g
Protein	1.8 g
Whey-protein	1.1 g
Casein	0.7 g
Lactose	2.4 g
Glucose	2.2 g
Dextrin-maltose	2.9 g
Minerals	0.4 g
kJ (Kcal)	315 (76)

Results

All results obtained from direct and indirect calorimetry are given in fig. 1. It can be seen that E(expenditure) measured by direct and indirect calorimetry increased linearly with increasing body weight. When expressed per kg bodyweight the results remain virtually constant.

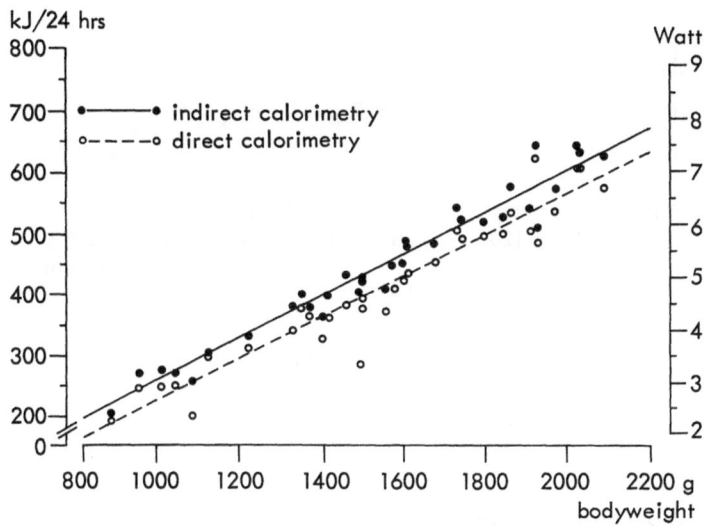

*Figure 1. Heat production as a function of
body weight*

E(components) was calculated from all the measurements using the energy balance equation:
E(in) = E(excreted) + E(expended) + E(components)
The results are given in fig. 2. E(components) was 11.9 ± 4 kJ/g growth. In all measurements the results of indirect calorimetry were higher than those of direct calorimetry. For the whole group the difference was on average 23 kJ/24 hr. or 7.2% of the indirect calorimetry. An explanation for this difference might be the following. Indirect calorimetry includes E(synthesis), as oxygen is consumed and carbon dioxide produced. Direct calorimetry on the other hand, measures heat loss and does not include E(synthesis) as this energy is stored in the body without

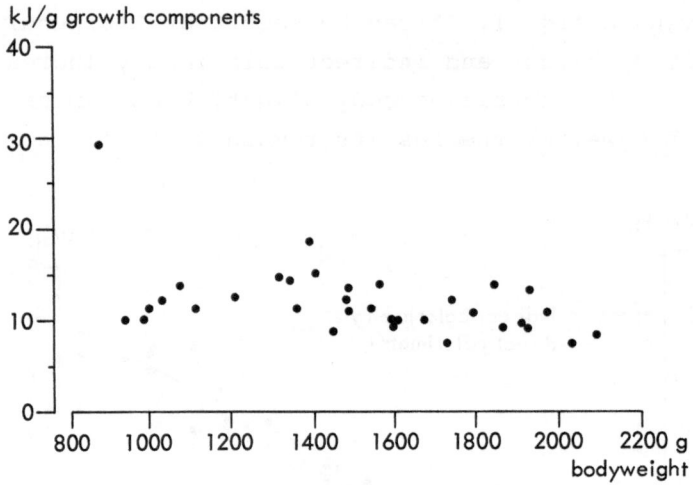

Figure 2. *Calculated energy cost of components of new tissue*

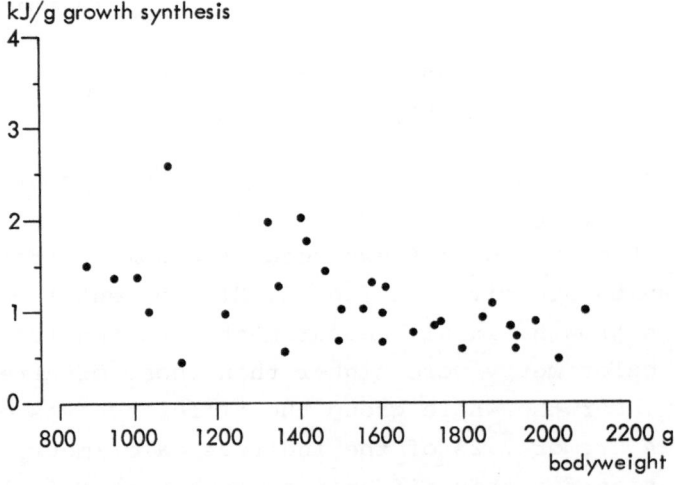

Figure 3. *Calculated energy cost of synthesis of new tissue*

increasing body temperature. Using this hypothesis we consider the difference between indirect and direct calorimetry as E(synthesis). E(synthesis) was calculated for all measurements and the results are given in fig. 3. E(synthesis) was 1.1 ± 0.5 kJ/g growth.

Energy balance

From our measurements it is possible to calculate the energy balance of a growing neonate. For a neonate of 1000 - 2000 g, growing 20 g/kg/24 hr., the energy balance (as kJ/kg/24 hr.) will be as follows (fig. 4): energy intake 525 kJ; energy loss in faeces and urine 55 kJ; energy used for maintenance, activity and thermoregulation 250 kJ; energy used for tissuesynthesis 20 kJ; and energy used for components of new tissue 200 kJ.

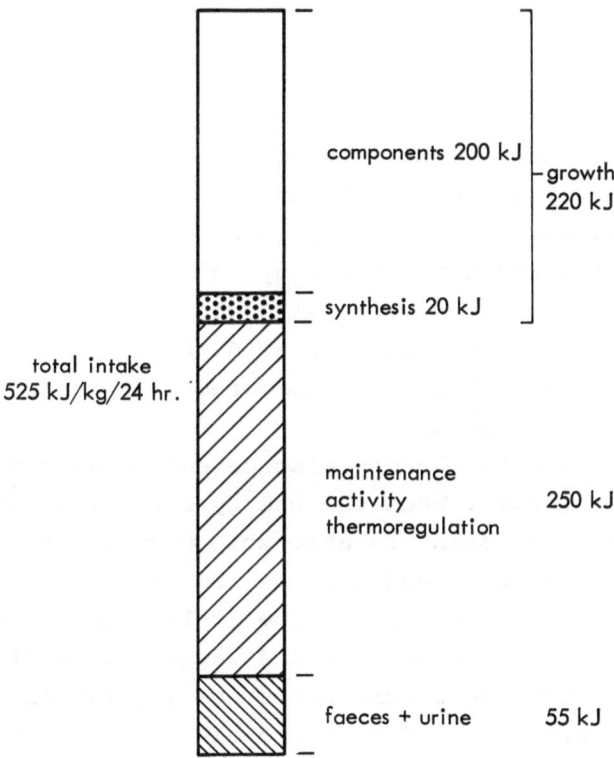

Figure 4. Energy balance of a neonate
growing 20 g/kg/24 hr.

Composition of newly formed tissue

Little is known about the composition of new tissue formed
during growth in the neonate. Calculations can be made
from carcass analysis, but a serious criticism of this
method is that these data are single measurements in
infants who had died and are not based on longitudinal
studies. Calculations of the composition of new tissue
made from data given in the literature on carcass analysis
are given in table 3.

Table 3. Composition of new tissue, data from carcass analysis

	±1000 - ±1500 gram		±1500 - ±2000 gram	
	protein g/g growth	fat	protein g/g growth	fat
Widdowson (1)	0.13	0.05	0.14	0.13
Ziegler (2)	0.11	0.10	0.12	0.12

In our hospital Mettau (14) has studied the fat content
of low birth weight infants longitudinally, using a Xenon-
absorption method. During this study these infants
received formula feeding with an energy intake of
± 580 kJ/kg/24 hr. The fat content of newly formed tissue
can be calculated from these studies as 0.16 g/g growth
(fig. 5). This figure is higher as compared with the data
from carcass analysis studies.

The composition of new tissue can also be estimated from
the energy cost of growth. When the energy cost of growth
is known, there is a constant relationship between the
amount of protein and fat layed down in new tissue. If the
energy cost of components of new tissue is 11.9 kJ/g growth
and if new tissue only contains fat and water, 31% will be
fat and 69% water assuming E(components) of 1 g fat is
37.6 kJ. On the other hand, if only protein and water is
layed down the composition of new tissue will be 71% protein and 29% water when E(components) of 1 g protein is
16.7 kJ. The amount of protein and fat in new tissue will

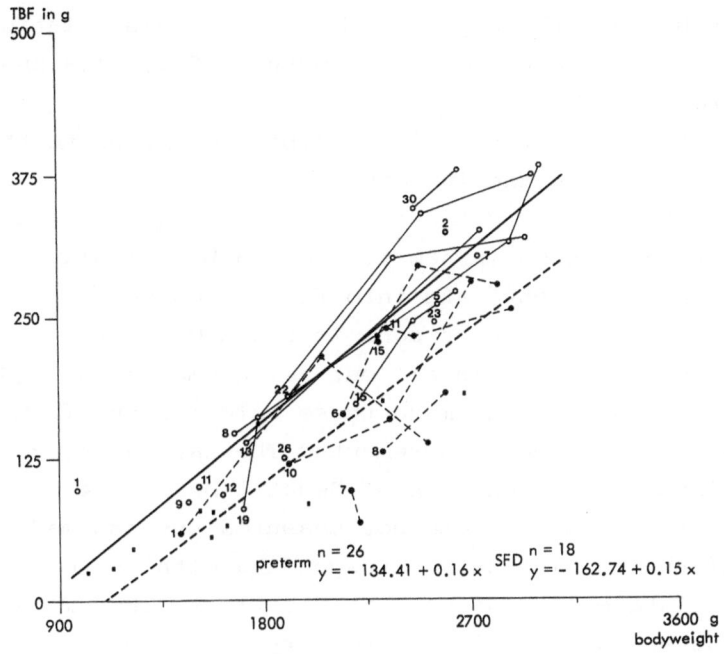

Figure 5. *Total body fat (TBF) plotted against body weight (BW). Regression lines are drawn for preterm babies (——) and for S.F.D. babies (---) separately. (data from Mettau (14))*

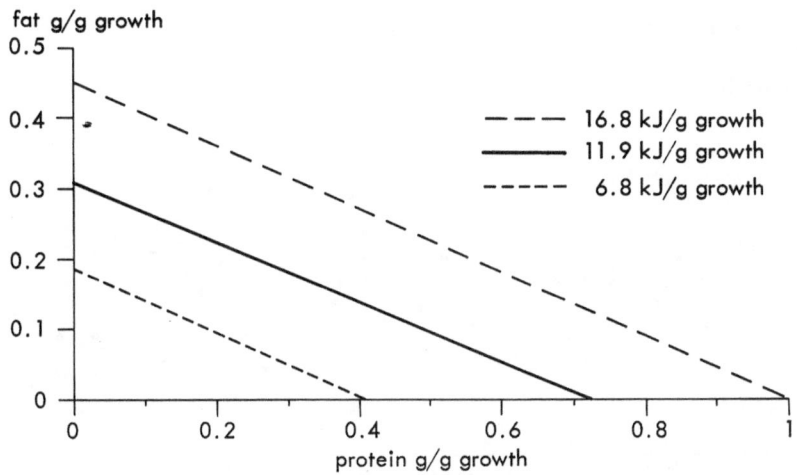

Figure 6. *Relationship between the amount of fat and protein layed down in new tissue*

be on a line between these points. In fig. 6 this relationship is given when energy cost of growth is 6.8, 11.9 and
16.8 kJ/g growth.

Using indirect calorimetry it is possible to calculate the
amount of carbohydrate and fat combusted. Knowing the fat
intake, the fat losses and the amount of fat oxidized, the
amount of fat layed down in new tissue can be calculated.
The conversion of carbohydrate into fat is neglected in
this calculation. This can be done if the R.Q. of the
measurement is below 1.0. In all our measurements the R.Q.
was between 0.9 and 1.0. We made these calculations from
the previously mentioned measurements. The fat intake was
calculated from the composition of Nenatal being 4.4 g/
100 ml. The faecal losses were not measured but assumed
to be 15% of the fat intake, corresponding with 10% of
total energy intake loss as mentioned earlier. The amount
of fat combusted was calculated from oxygen consumption
and carbon dioxide production.

We calculated the amount of fat per gram of new tissue as
0.10 - 0.34 g fat/g new tissue (fig. 7). So it would seem
that the composition of new tissue was fairly constant
during growth from 1000 to 2000 gram in this group of small
for date infants.

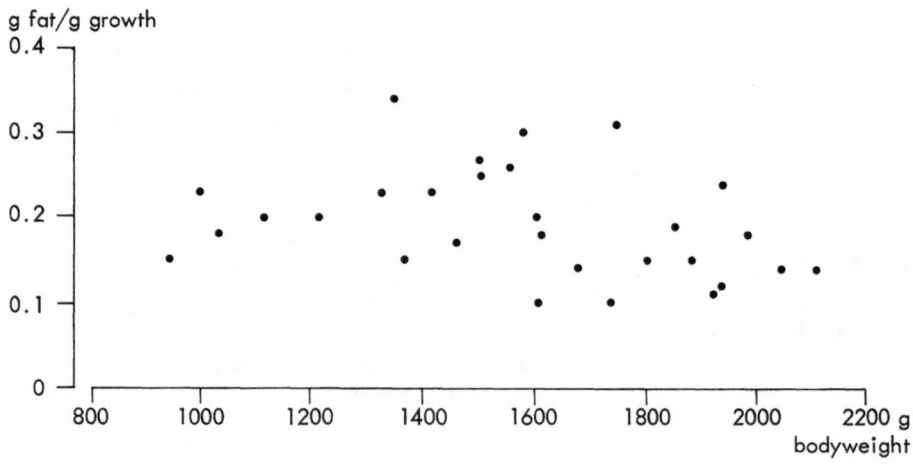

*Figure 7. Fat content of new tissue plotted against body
weight*

Summary

The energy requirements for growth consist of two parts,
the energy present in the components of new tissue and
the energy necessary to organize the new tissue. Using a
new calorimeter in which it is possible to do direct and
indirect calorimetry continuously and simultaneously for
several hours, we measured the energy balance in 6 small
for dates infants weighing between 870 and 2100 g at the
age of 8 to 58 days. Thirty-four studies were made.
In all measurements the results of indirect calorimetry
were higher than those of direct calorimetry. The difference is considered to be the energy cost of tissue
synthesis (1.1 ± 0.5 kJ/g growth). The energy used for
components of new tissue is calculated as 11.9 ± 4 kJ/g
growth. From our data it was possible to estimate the
fat content of new tissue as $0.10 - 0.34$ g/g growth.

References

1. Widdowson, EM, Changes in body proportions and composition during growth. In: Scientific Foundations of
   Paediatrics, Davis, JA and J Dobbing (eds), London,
   Heinemann, 153-163, 1974
2. Ziegler, EE, AM O'Donnell, SE Nelson and SJ Fomon,
   Body composition of the reference fetus. Growth 40:
   329-341, 1976
3. Sinclair,JC, Energy balance of the newborn. In: Growth
   and development of the full-term and premature infant,
   Jonxis, JHP (ed), Amsterdam, Excerpta Medica, 19-40,
   1978
4. Sauer, PJJ, RG Pearse, HJ Dane and HKA Visser,
   The energy cost of growth estimated from simultaneous
   direct and indirect calorimetry in infants of less
   than 2500 g. In: Nutrition and metabolism of the fetus
   and infant, Visser, HKA (ed), The Hague, Martinus
   Nijhoff, 93-107, 1979
5. Brooke, OG, J Alvear and M Arnold, Energy retention,
   energy expenditure, and growth in healthy immature
   infants. Pediatr Res 13: 215-220, 1979

6. Ashworth, A, Growth rates in children recovering from protein-calorie malnutrition. Br J Nutr 23: 835-845, 1969

7. Spady, DW, PR Payne, D Picou and JC Waterlow, Energy balance during recovery from malnutrition. Am J Clin Nutr 29: 1073-1078, 1976

8. Howland, J, Direct calorimetry of infants, with a comparison of the results obtained by this and other methods. Tr 15th Internatl Cong Hyg and Demog ii: 438, 1912

9. Day, R and JD Hardy, Respiratory metabolism in infancy and in childhood. Am J Dis Child 63: 1086-1095, 1942

10. Day, R, J Curtis and M Kelly, Respiratory metabolism in infancy and childhood. Regulation of body temperature of premature infants. Am J Dis Child 65: 376-398, 1943

11. Brooke, OG and A Ashworth, Influence of malnutrition on the postprandial metabolic rate and respiratory quotient. Br J Nutr 27: 407-415, 1972

12. Hommes, FA, YM Drost, WXM Geraerts and MAA Reijenga, The energy requirements for growth: An application of Atkinson's metabolic price system. Pediatr Res 9: 51-55, 1975

13. Usher, R and F McLean, Intrauterine growth of live-born Caucasian infants at sea level: Standards obtained from measurements in 7 dimensions of infants born between 25 and 44 weeks of gestation. J Pediatr 74: 901-910, 1969

14. Mettau, JW, Measurement of total body fat in low birth weight infants. Thesis, Rotterdam, 1978

15. Kielanowski, J, Estimates of the energy cost of protein desposition in growing animals. In: Energy Metabolism, Blaxter, KL (ed), London, Academic Press, 13-20, 1965

DISCUSSION

Qu.Teller : I have a problem understanding your expression
"g/growth". What is its exact definition, because there is
weight gain without growth and growth without weight gain?
To my mind growth means increase in length. Is this correct?
An.Sauer : We have expressed our results as energy cost of
growth per gram growth. With growth we have meant increase
in body mass. I completely agree with you that there can
be weight gain without increase in body mass and increase
in body mass without weight gain. However, during the
period of our study, the infants showed a continuous increase in weight. We believed we could consider this increase
in weight as an increase in body mass, what we have called
growth.

Com.Moore : The data is very impressive and it is good to
see the comparison between direct and indirect calorimetry
so beautifully carried out. The R.Q.'s were high (0.9 to
1.0) and is presumably related to the high carbohydrate
content of this diet, and also to the conversion of carbohydrate to fat.
An.Sauer : The R.Q.'s of our measurements are indeed quite
high. It is possible to calculate from the amount of oxygen consumed and carbondioxyde produced, the amount of carbohydrate and fat combusted during the measurement period.
We calculated that in all our measurements nearly all carbohydrate and a part of the fat of the feeding given was
combusted. So I believe the high R.Q. is due to the amount
of carbohydrate used by the neonates and not to the conversion of carbohydrate to fat.

Qu.Britton : From your calculations it appears that the
energy cost of growth is very large in comparison to energy
stored. Could this be related to the caloric/nitrogen ratio
in the food? In other words, may the infant simply be for-
ced to burn off the excess calories in his diet? If this is
the case, the proportion of ingested nitrogen converted to
protein might be expected to be high. Have you any evidence

about this, for example from urinary nitrogen excretion?

An.Sauer : During this study we did not measure urinary
nitrogen excretion. A large part of the energy given with
food is used for growth. However, during this study the
infants were growing rather fast. The energy cost of growth
expressed per gram growth is not high, lower than known from
other studies. From other studies, it became clear that
when the energy used for thermoregulation is increased, the
growth rate is decreased. When the caloric intake then is
increased, growth rate will also increase. How far growth
can be accelerated by an increase in caloric intake is not
known.

Qu.Senterre : I have some difficulties to understand your
approach of the problem because, in my opinion, direct and
indirect calorimetry must give the same results. The differences you measured, perhaps may be interpreted as high
energy bounds inside the molecule. However most of energy
cost of growth is dissipated as heat and measured by an
increase in O_2 consumption with both methods. Energy cost
of growth may be related to the increase in basal metabolic
rate when the infant is growing. But what you measured is
not the real energy cost of growth. Can you comment about
this point of view?

An.Sauer : Energy cost of growth consists of two parts energy present in the components of new tissue and energy necessary to organize the components into the new tissue.
The latter might for instance be the high energy bounds inside the molecules. This energy is stored in the body together with the energy present in the components of the new
tissue and will not be given off as heat, so it cannot be
measured with direct calorimetry. On the other hand, foodstuffs are combusted for the conversion of this energy. So
this energy is measured by indirect calorimetry. It might
be that the increase in metabolic rate measured with indirect calorimetry when the infant is growing, is related to
the energy cost of growth. However, we realize that our
model is too simple as the process of tissue synthesis is

not 100 % efficient, like all processes in the human organism. When we believe the processes in the human body have
an efficiency of 50 %, the total energy cost of tissue synthesis will be 2,2 kJ/g growth instead of the 1,1 kJ/g
growth I mentioned earlier. Of this 2,2 kJ, half of it will
be given off as heat and be measured by both direct and indirect calorimetry. So we believe the difference between
direct and indirect calorimetry is that part of the energy
cost of tissue synthesis that is stored in the body.

THE ROLE OF OLEIC ACID AS A SUBSTITUTE FOR GLUCOSE IN BRAIN CELL CULTURES OF NEONATAL MICE

Emilio Bossi, M.D. and Hans Peter Siegrist, Ph.D.

The effects which most of the substances utilized in neo-
natology may exert on the brain of the newborn are unknown.
For example, there is a lack of basic knowledge on the
effects of the components of parenteral nutrition (amino
acids, fat and even glucose) on the developing brain. By
using dissociated brain cell cultures of newborn mice as a
model for studying brain metabolism, we investigated
certain aspects of cerebral energy metabolism. One aspect
we decided to study was the role of alternative energetic
substrates in brain during glucose deprivation.

Methodology of cell cultivation

This method has been elaborated by Dr. Ulrich Wiesmann in
collaboration with Dr. Karel Hoffmann in our laboratories
(4). It will be shortly summarized. A cell suspension is
prepared from brains of newborn mice within 12 hours of
birth, one brain yielding cells for 2 culture flasks. The
cells are then grown in Dulbecco medium (modification
DMEM) containing 10 % fetal calf serum (Seromed, Munich)
and 27 mM glucose. The medium is changed every third day.
To induce glucose deprivation, media containing different
concentrations of glucose are substituted for the culture
medium at day 13 to 15.
The cultures consist mainly of astrocytes, of oligodendrocytes (20 %) and of a basal layer of unidentified cells.
They contain practically no neurons, since their replication in mouse brain is completed at the time of birth and

culture is therefore not possible. We would like to point out that, by this model, metabolic processes are studied directly on the cell. Thus the influence of the blood-brain barrier and generalised metabolic effects can be excluded.

Glucose sensitivity of the neonatal brain cell cultures

In order to study the role of substitutes for glucose as fuels, we had first to show that our cell model was sensitive to glucose deprivation. Exposure to low glucose concentrations leads to morphologic changes of certain cells, as identifiable by light microscopy. By producing anti-oligodendrocytes- and anti-cerebroside-antibodies in rabbits, Dr. Liane Sandru and Prof. Norbert Herschkowitz of our group have shown these cells to be oligodendroglia. Thus, a specific function of oligodendrocytes, sulfatide synthesis, could be measured at different glucose concentrations. For this purpose, Dr. Klaus Zuppinger performed the following experiments: the cells were exposed to 0.8, 2.8, 4.8, 7.3, 13.6 and 25.2 mM glucose at day 10 and left for 72 hours in these media. After 48 hours, 200 μCi $^{35}SO_4$ were added. 24 hours later, total lipid extraction was carried out and $^{35}SO_4$ counted. Sulfatide synthesis was defined as dpm $^{35}SO_4$ per mg protein. A glucose dependency of sulfatide synthesis could clearly be demonstrated (Figure 1).

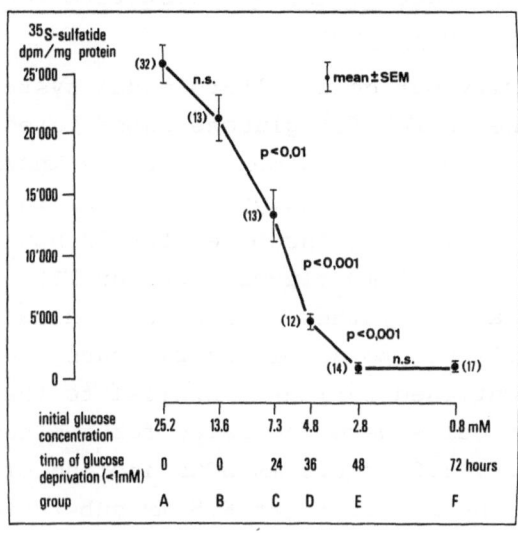

Figure 1   Glucose dependency of sulfatide synthesis
            in dissociated brain cell cultures of
            newborn mice

The activity of a specific oligodendrocyte enzyme,
cerebroside sulfotransferase, was also shown to be glucose
dependent (Table I).

Table I   Effect of glucose on cellular cerebroside
            Sulfotransferase activity

Group	Initial glucose	Cerebroside sulfotransferase activity dpm ^{35}S-sulfatide per mg protein $\bar{x}$ ± SEM		Sulfatide synthesis as ^{35}S incorporation into sulfatide
	mM		%	%
A	25.2	24'894 ± 1'676 (13)	100	100
B	13.2	27'175 ± 2'968 (7)	109.2	82.4
C	7.3	7'288 ± 1'557 (5)	29.3	51.3
D	4.8	3'823 ± 445 (8)	15.4	17.9
E	2.8	2'194 ± 342 (5)	8.8	3.5
F	0.8	2'561 ± 352 (9)	10.3	3.8

D-Betahydroxy-Butyrate (BOHB) as substitute for glucose in promoting sulfatide synthesis

In order to verify our cell culture model system as suitable for the search for glucose substitutes as cerebral energetic fuels, we had to demonstrate the known effect of ketone bodies as cerebral energy substrates in our system. (1, 2, 3). The cells were incubated for 72 hours with either 2.1, 4.2 or 6.5 mM glucose. On day 13, $^{35}SO_4$ was added, and sulfatide synthesis from the radioactive precursor was allowed to go on for 24 hours, when it was estimated as mentioned earlier. Parallel to these cultures, other cells were incubated for the same period of time with 2.1, 4.2 or 6.5 mM BOHB instead of glucose. Figure 2 shows the results: at 6.5 mM substrate concentration, sulfatide synthesis was better sustained by glucose. At 4.2 mM, both glucose and BOHB had the same effect, although on a lower level. 2.1 mM BOHB, however, was even more effective than 2.1 mM glucose in sustaining sulfatide synthesis (5). Thus, the role of BOHB in cerebral energy metabolism could be shown in our model.

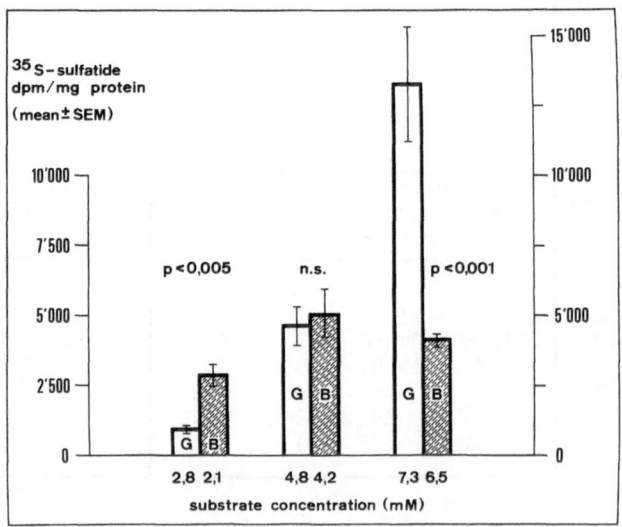

Figure 2   Sulfatide synthesis at different concentrations of either glucose (G) or D-Beta-OHButyrate (B) in the same model as in figure 1.

Utilization of BOHB as an energetic substrate

We then asked the question if BOHB was really itself utilized as an energetic fuel. We answered this question by adding D-Beta-Hydroxy-Butyrate-3-^{14}C and measuring $^{14}CO_2$-production. Should $^{14}CO_2$ be produced from the labelled compound, this would be the proof for BOHB-metabolization in the Krebs-cycle and thus for a role of the ketone body as an energetic substrate.

The following <u>methodology</u> was developed, as illustrated in <u>Figure 3</u>.

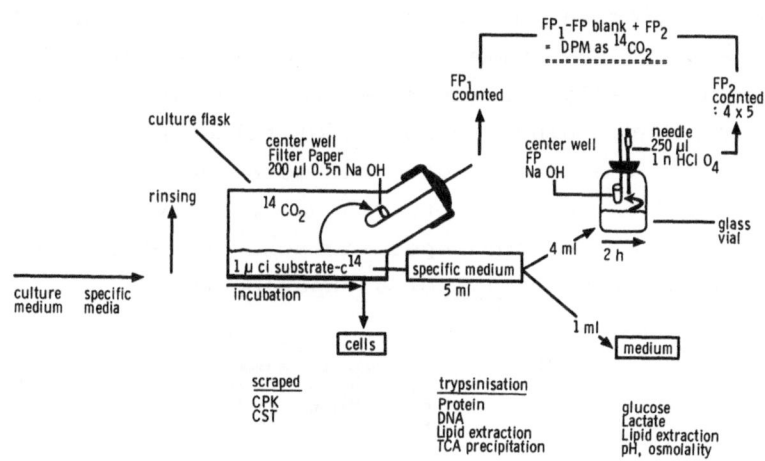

Figure 3   Collection of $^{14}CO_2$ produced by brain cells in culture from ^{14}C labelled energy substrates

The cells were cultivated for 13 days in the culture medium. Then, the medium without glucose was added. After 48 hours, the cells were rinsed with glucose-free medium and 5 ml medium containing 4 mM D-BOHB and 1 μCi D-BOHB-3-^{14}C and either 0,6.5 or 27 mM glucose was added. The flask was closed with a screw cap equipped with a plastic center well (Mercatura, Basel) containing a filter paper (Whatman glassfiber papers GF/A 2.4 cm) moistened with 200 μl 0.5 n NaOH. (Flasks: Tissue Culture flask 25 cm^2, canted neck,

Falcon). Incubation was allowed to proceed for 1/2 to 4 hours. The $^{14}CO_2$ produced from BOHB-3-^{14}C by the cells was trapped on the NaOH-moistened filter paper. After the incubation, the filter paper was transferred into a counting vial and 10 ml scintillation fluid (Dimilume^{30}, Packard) were added. 4 ml of the medium were transferred into a glass vial. This was then tightly closed with a rubber stopper, through which a central well containing a NaOH-moistened filter paper and an injection needle had been introduced. Through the needle, 250 μl 1 n $HClO_4$ were injected in order to drive off the $^{14}CO_2$ remaining in the medium after the incubation period. The cells were either scraped off (for enzyme determinations) or trypsinized and further analysed as summarized in the figure. After 2 hours, the second filter paper was taken for counting. The counts of filter paper 1 plus the counts of filter paper 2, corrected for the blanks, gave the total dpm as $^{14}CO_2$. BOHB-utilization in μg/mg protein was finally calculated on the basis of the known specific activity of D-BOHB-3-^{14}C. Figure 4 shows the results of 6 experiments.

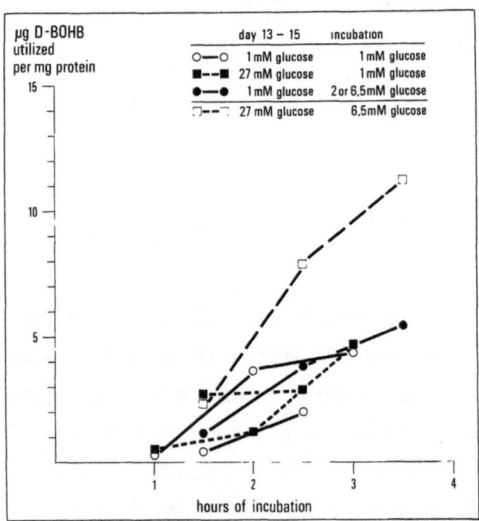

Figure 4   D-BOHB utilization by cells exposed to low glucose concentrations for certain time periods vs. cells constantly offered higher glucose concentrations. Standard deviations (not shown) ranged mainly within 10 % and never exceeded 20 %.

The experiments give evidence that BOHB is metabolized to CO_2. The time curve is almost linear. Thus, BOHB is utilized as a fuel for brain cells in culture. In addition, cultures which had never been exposed to less than 2.0 mM glucose utilized clearly more BOHB than those which had been exposed to very low glucose concentrations. This phenomenon is probably due to a better viability of the cells exposed to higher glucose levels.

Utilization of fatty acids (FA) by the brain cell cultures

Fatty acid concentrations in the media were determined by gas chromatography. According to the pattern in fetal calf serum, the following relative concentrations were found (Table II).

Table II   Relative concentrations of the main fatty acids in the incubation medium. These fatty acids are present as free fatty acid pool and esterified with phosphatidyl-glycerol and cholesterol

Linolic acid	5	%
Arachidonic acid	8,8	%
Stearic acid	16	%
Palmitic acid	20	%
Oleic acid	38	%

In order to investigate the fate of fatty acids in the media, we determined them before and after the glucose deprivation of cells in culture in a same experimental set up as was used for the preceeding experiments. The free concentrations of FA were determined at the beginning and after 48 hours of incubation at different glucose concentrations. In Figure 5, a typical pattern can be seen for the FA examined.

216

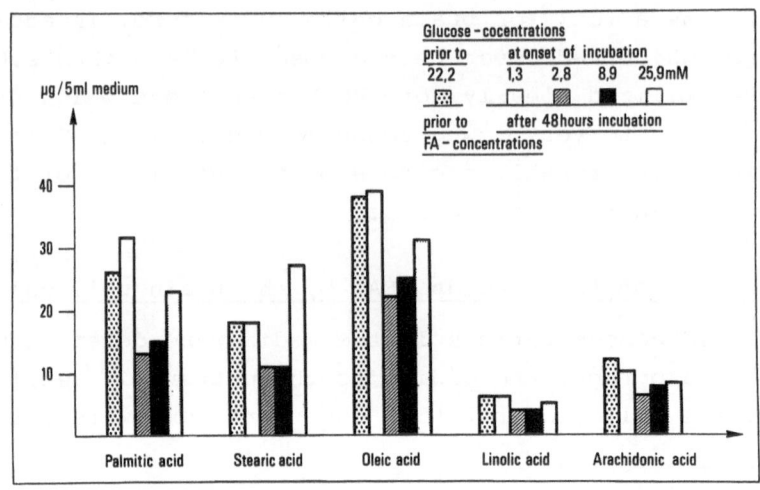

Figure 5   Concentrations of FA prior to and after 48
            hours incubation in culture media of neonatal
            mouse brain cells at different glucose con-
            centrations. (1 representative experiment
            out of 4)

At glucose levels of 25.9 mM, no difference between the
initial and final FA content of the medium is found. The
lowest FA concentrations after 48 hours are found at 2.8
mM glucose, indicating utilization of the FA by the cells.
The more the glucose concentration is increased, the less
FA disappear from the medium. Thus, the FA examined (which
amount to 87.8 % of the total FA in the medium) seem to be
utilized by the cells. This utilization is dependent on
the glucose concentrations offered to the cells and is
most pronounced at 2.8 mM glucose. We interpret the fact
that no utilization occurs at 1.3 mM glucose again as a
consequence of a diminished viability of the cells at
such low glucose levels. Another hypothesis for explaining
this phenomenon would be that a certain minimal concentration of glucose is required to activate the Krebs-cycle.

Utilization of oleic acid as an energetic substrate

As for BOHB, we wanted to know whether FA are utilized in
the energy metabolism. Therefore, 1 μCi oleic acid-UL-^{14}C
(which is the main component of the fatty acid pool) was
added to the medium and $^{14}CO_2$ production measured as
described above for D-BOHB-3-^{14}C. In <u>Figure 6</u>, where the
results of 3 experiments were pooled, a linear kinetic of
$^{14}CO_2$ production can be observed.

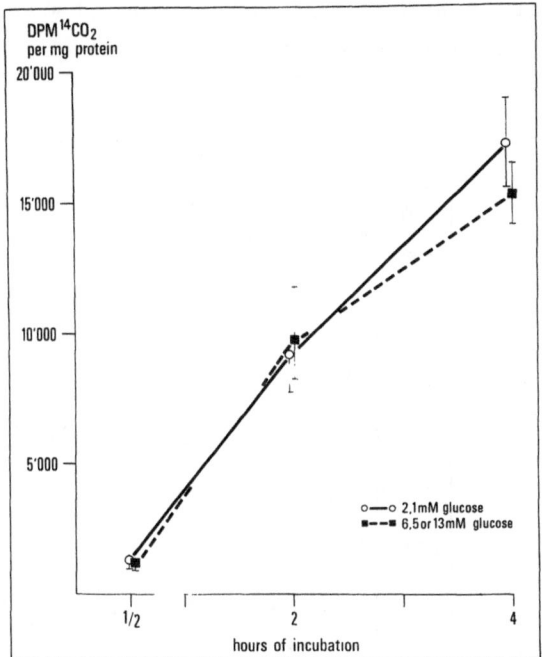

<u>Figure 6</u> $^{14}CO_2$ produced from oleic-acid-UL-^{14}C by
brain cells in culture, exposed to
specific media during incubation and for
2 days before. Mean $\pm$ 1 SD

Thus, oleic acid is degraded to $^{14}CO_2$ by brain cells in
culture. No difference is noticed between the $^{14}CO_2$ pro-

duction of cells exposed to different glucose concentrations.

For the further experiments, the medium was dialysed in
order to eliminate as much glucose as possible. We wanted
to see if a higher utilization of oleic acid could be found
at extremely low glucose concentrations. In the experiments
with BOHB, no cells had been exposed to strictly no
glucose. Even when no glucose was added, the medium contained some glucose (approximately 1 mM) contributed by the
fetal calf serum. $^{14}CO_2$-production from oleic acid was
determined after 4 hours of incubation at different glucose
concentrations. Figure 7 shows the results of the 2 first
experiments.

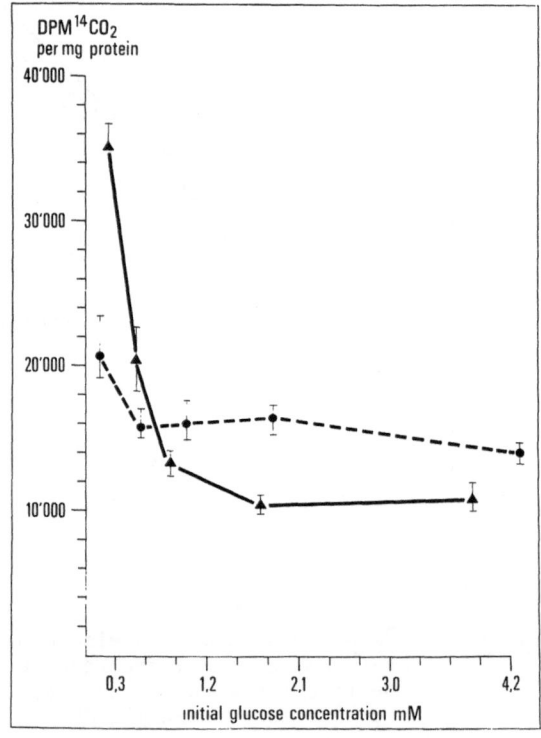

Figure 7   Production of $^{14}CO_2$ from oleic acid-UL-^{14}C
             by neonatal mouse brain cells after 15 days
             in culture, at different glucose concentra-
             tions during a 4 hours incubation period.
             (The ranges are indicated).

In fact, significantly more oleic acid is utilized for
energy production at the extremely low glucose concentrations now achieved (0.15 - 0.5 mM) than at glucose concentrations above 0.6 mM.

Summary

Cultures of dissociated neonatal mouse brain cells, con-
sisting mainly of astroglia and oligodendrocytes, were
found to be glucose sensitive with respect to sulfatide
synthesis. D-BOHB substitutes for glucose in promoting
sulfatide synthesis. Its utilization for energy production
was proven by showing its conversion into $^{14}CO_2$ when
D-BOHB-3-^{14}C was used. Furthermore, the concentration of
the main FA contained in the medium diminished during a
48 hours incubation. The most pronunced reduction was
found at 2.8 mM glucose; at higher or lower glucose concentrations, disappearance from the medium was reduced,
indicating that fatty acids are taken up by the cells at
low glucose concentrations, but that some glucose is still
necessary for this to occur. When oleic acid-UL-^{14}C was
added to the medium, it was metabolized to $^{14}CO_2$ by the
cells, indicating its utilization as fuel. At glucose concentrations below 0.6 mM, as compared to higher glucose
concentrations, significantly more oleic acid was used as
an energy substrate.

Outlook

The latter experiments presented are to be considered as
preliminary. Further confirmation of the experiments with
oleic acid is necessary. Conversion of BOHB to CO_2 is
going to be investigated at lower glucose concentrations
by using dialysed serum. Other possible substrates for
energy production by neonatal brain cells will be investigated, as for instance glycerol. Furthermore, we plan to
study developmental aspects of the role of substitute
energy substrates.

References

(1)   Adam P.A.J.; Räihä N.; Rahiala E-L.; Kekomäki M.:
      Oxydation of glucose and D-B-OH-Butyrate by the
      early human fetal brain.
      Acta Paediatr. Scand. 64: 17 (1975).

(2)   Kraus H.; Schlenker S.; Schwedesky D.:
      Developmental changes of cerebral ketone body
      utilization in human infants.
      Hoppe-Seyler's Z. Physiol. Chem. 355: 164 (1974).

(3)   Owen O.E.; Morgan A.P.; Kemp H.G.; Sullivan J.M.;
      Herrera M.G.; Cahill G.F.Jr.:
      Brain metabolism during fasting.
      J. Clin. Invest. 46: 1589 (1967).

(4)   Wiesmann U.N.; Hofmann K.; Burkhart T.; Herschkowitz
      N.:
      Dissociated cultures of newborn mouse brain.
      Neurobiology 5: 305 (1975).

(5)   Zuppinger K.; Bossi E.; Wiesmann U.; Hofmann K.;
      Schaefer T.; Herschkowitz N.:
      Dependency of sulfatide synthesis upon glucose and
      D-BetaOH-Butyrate concentration in the medium of
      dissociated mouse brain cell cultures.
      Pediat. Res. 13: 84 (1979), Abstract.

Acknowledgement: The authors are very grateful to
Mrs. Ursula Wanzenried and Miss Therese Schaefer for
their excellent technical collaboration.

This work was supported by Schweizerischer Nationalfonds
zur Förderung der wissenschaftlichen Forschung, grant no.
3.972-0.78.

DISCUSSION

<u>Qu.Ferre</u> : Are the cells, after 13 days of culture, representative of real brain composition?

<u>An.Bossi</u> : No. The ratio neurons-oligodentrocytes is changed in favour of the oligodendrocytes.

<u>Qu.Roux</u> : Do you have cell multiplication?

<u>An.Bossi</u> : Yes.

<u>Qu.Milner</u> : Was the fetal calf serum heated or unheated?

<u>An.Bossi</u> : Unheated.

<u>Qu.Milner</u> : What happens if fetal calf serum is omitted - throughout - from day 13?

<u>An.Bossi</u> : We know that cell survive up to 72 hours in serum-free medium. Therefore, our experiments cannot be carried out in such a system.

<u>Qu.Helge</u> :Did you have a chance to look for the effects of β- hydroxybutyrate and fatty acids on the growth and multiplication of the cultured cells? What happens, when they are exposed from the beginning and for longer time periods, not just for 48 hours at day 15 of the culture?

<u>An.Bossi</u> : This kind of experiment has not been carried out. Our system is based on short term experiments in order to measure CO_2-production from a certain substrate which would be metabolized in long term experiments. The study of this relevant question would require another experimental design.

<u>Qu.Britton</u> : As a minor question of detail can you be sure that all of the radioactivity derived from 3- $[^{14}C]$ β hydroxybutyrate was in the form of $^{14}CO_2$? Decarboxylation of acetoacetate could yield acetone which would be absorbed by the alkali. Have you calculated the amount of β hydroxybutyrate oxidised and is this quantity sufficient to make a substantial contribution to the oxidative metabolism of the cell? In muscle, β hydroxybutyrate and free fatty acids have a sparing effect upon glucose metabolism. Was there any

reduction in glucose uptake in your preparation in the
presence of these substrates?

An.Bossi : Treatment of the filter papers with HCl drived
off all of the radioactivity. This would not occur if part
of the radioactivity would be acetone.

Our type of cultures is a good model for studying metabolic
pathways. Since the distribution of cell does not correspond to the in vivo situation, however, quantitative comparisons with whole organ systems cannot be carried out.

Since no experiments with different concentrations of cold
substrates have been carried out, we cannot answer this
question. All we know is that, during the 4 hours incubations, the glucose glucose metabolism concentration in the
media diminishes.

Qu.Moses : Have you studied older animals?
An.Bossi : Older animals have not been studied.

Qu.Moses : Have you any data or correlation between in vivo
and brain cell culture studies of this type? Namely β
hydroxybutyrate utilisation and oleic acid metabolism by
the brain?

An.Bossi : Since we are at the beginning of our experimental
work, the data you ask for are not available.

DISCUSSION ABOUT THE SECOND SESSION

Qu.Britton : McCance emphasized how the composition and
quantity of milk in different species was closely related
to the particular calorific and nitrogen requirements of
the newborn of that species. In the human, the milk is presumably appropriate for the rearing of the newborn at below
the thermoneutral temperature. Can we therefore justify
the use of a thermoneutral temperature in incubators when
the babies are fed on breast milk?
Is it possible that such infants may receive a relative surfeit of calories?

An.Hull : I think we would only use incubators to nurse
sick newborn infants and preterm infants who are not thri-
ving. I accept the implication in your question that it is
not self evident that it is desirable to nurse healthy preterm infants in constant thermo-neutral conditions. Preterm infants do not grow as fast as we would expect them to
do if they had not been born and continued to grow in utero
so it is difficult to conclude that currently they receive
a surfeit of calories.

Qu.Girard : What is in vivo the quantitative significance
of FFA in brain oxydative metabolism?
An.Bossi : The quantitative significance is unknown.

Qu. Helge : May I ask you for an explanation or speculation
concerning the differences in N-retention and fat deposition after feeding breast milk or a formula diet to the
premature infants you described ?
Apparently, it was not the total caloric intake or overall
composition of the milk with respect to total nitrogen or
fat content. Can you tell us more about the amino acid or
fatty acid components of the breast milk and formula used ?
Did you try to change the formula in order to better adapt
it to the breast milk and there by influence N- and fat-
retention ?
An. Verellen : Composition of the milk with respect to total nitrogen is relevant. For a same net energy intake of

117 Kcal/kg/day, breast milk fed babies were receiving respectively 2,9 - 5,6 and 13,8 g/kg/day of metabolisable
protein, fat and carbohydrate. Formula fed babies were receiving respectively 1,8 - 5,7 and 14,7 g/kg/day. This does
indicate that at a same net energy intake, nitrogen intake
was higher in the infants fed their own mother's early
milk. This may explain the higher nitrogen retention observed in this group. Other possible causes are a more favora-
ble amino acid profile and a large fraction (25 %) of non
protein nitrogen.
We did not try to modify the formula.

Com. Minkowsky : The reference to intrauterine growth for
post-natal growth does not seem to be justified.
An. Verellen : Intra-uterine growth rates as the ideal
towards which one should work in the nutritional management
of the premature is still a reasonable goal for many neo-
natologists. We agree that this quantitative objective is
not proved to be the best one and that a rational nutritional design should also be based on the accretion rates of
the principal nutrients during intra-uterine life. The
present study demonstrates that quality of growth (protein
versus fat deposition) in response to a specific diet composition may be defined and hopefully will provide guidelines
for the quantity and quality of nutrients for the optimal
extra-uterine development of the very low birth weight infants
during the first postnatal weeks.

SESSION III
REGULATION OF INSULIN AND
GLUCAGON SECRETION

INSULIN SECRETION AND METABOLISM IN THE RAT FETUS DURING LATE GESTATION

F. Sodoyez-Goffaux, J.C. Sodoyez, C.J. de Vos, and Y.M. Thiry-Moris

In most species, the synthesis and storage of insulin in the pancreatic B cells begins at an early stage of intra-uterine life. In the human fetus, differentiated B cells may be recognized at about 10.5 weeks of gestation (1). In the rat, immunoreactive insulin (IRI) is first detected on the 11th day (2) and the amount of pancreatic IRI available per g body weight briskly increases during the last few days of a 22 day gestation (3).

To evaluate the secretory capacity of fetal and neonatal pancreases, we measured the insulin output of isolated islets incubated in a medium containing 50 mg % glucose (a), the insulin content per islet (b) and the total insulin content of the pancreas (c) and, using the equation $\frac{a}{b} \times c$, calculated the total secretory capacity of the pancreas. To take body growth into account, the total secretory capacity of the pancreas was divided by the body weight. By doing so, we observed that the secretory capacity per unit body weight increased from .6 mU/h/g 2 days before term to 1.5 mU/h/g 1 day after birth.

Increasing insulin stores and secretory capacity contrast with the well known perinatal profile of plasma IRI concentration characterized by hyperinsulinemia before birth and a brisk postnatal decrease of plasma IRI concentration (4). This observation implies that, in vivo, insulin secretion is already modulated either by stimulating factors operating as long as the fetuses are in utero or by inhibiting factors operating after birth or by a combination of the 2 mechanisms. The alternate hypothesis

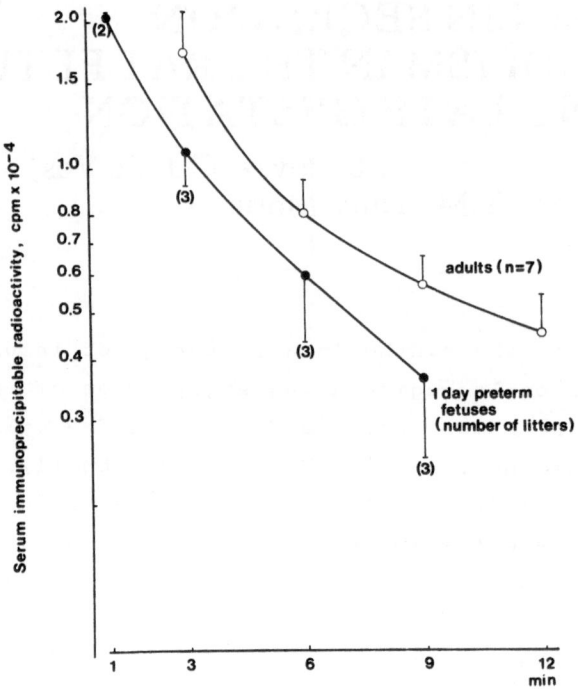

Figure 1: Disappearance rate of serum immunoprecipitable radio-activity after I V bolus injection of ^{125}I-insulin into 1 day preterm fetuses and adult rats. Ave $\pm$ SD.

that fetal hyperinsulinemia merely results from insulin accumulation in the blood because of slow uptake by the fetal tissues was ruled out by the observation that the plasma clearance rate of I.V. injected ^{125}I-insulin was at least as high in 1 day preterm fetuses as in adult rats (figure 1).

The distinctive characteristic of adult B cell is its ability to respond to minute to minute changes in the organism's supply of fuels by an appropriate change in insulin secretion. Although questioned for many years, the sensitivity of the fetal B cells to glucose has now been well demonstrated in vivo (5) and in vitro (6). At all the tested ages and already on the 19th day of gestation, the earliest date at which islets could be harvested, glucose appeared an effective stimulus of

insulin secretion. When compared to that of adult B cells,
the response of fetal islet was however not optimal.
During the last days of gestation, the circadian profile
of maternal blood glucose is fairly low and flat. Fetal
blood glucose is even lower, never reaching stimulatory
levels. Moreover, fetal blood glucose and plasma IRI
concentrations do not correlate well. These considerations
do not support the hypothesis that, under normal conditions,
glucose plays an important role in the regulation of fetal
insulin secretion in vivo.

Aminoacids have long been known to stimulate the
insulin release of fetal B cells (7,8) and interestingly,
a high plasma aminoacids concentration prevails during
late fetal life. Aminoacids may therefore be, at least
in part, responsible for fetal hyperinsulinemia.

Innumerable factors other than fuels are susceptible
to modulate insulin secretion. A possible role of cephalic
factors has been suggested by the observation of Van Assche
et al (9) that anencephalic newborns of diabetic mothers
develop hyperinsulinemia only if their hypothalamo-
pituitary system is intact. Islet stimulating factors
have been extracted from the hypothalamus (10) and from
the pituitary gland (corticotrophin like intermediate lobe
peptide or CLIP) (11). Yet, the hypothalamo-pituitary
axis does not appear necessary for the normal development
and histological differentiation of the fetal endocrine
pancreas (12). The fetal B cells are already reactive to
neurotransmitters in vivo (13) and the autonomic nervous
system may indeed contribute to the modulation of insulin
secretion, particularly at birth. Finally, a possible
role of gastrointestinal factors has also been proposed.
Factors such as gastric inhibitory peptide (GIP) have
been shown to stimulate the insulin output of neonatal rat
pancreases (14) and it is conceivable that the secretion
of these gastrointestinal factors might be triggered by
the swallowing of amniotic fluid. Decapitation in utero,
however, a condition which obviously prevents the

swallowing of amniotic fluid, did not modify plasma IRI
level in the rat fetus (15).

During the past years, we gathered experimental
evidences suggesting that the placenta itself modulated
fetal insulin secretion. If rat pups were delivered by
caesarian section one day before term, their plasma IRI
concentration rapidly decreased and adjusted to the same
low level as in pups spontaneously delivered at term.
Thus, birth itself and not some kind of biological clock
triggers the postnatal decrease in insulin secretion.
As witnessed by the catecholamine (16) and glucagon surge
(17), birth is of course a stressful condition. However,
if we minimized the stress of birth by transferring the
fetuses immediately after caesarian section into a well
humidified and oxygenated incubator at 37°, we observed
no glucagon or corticosterone surge and yet, plasma IRI
concentration abruptly fell. Moreover if the postnatal
insulin secretion switch off was due to inhibiting factors,
these should persist throughout extrauterine life since
the plasma IRI levels will never return to the record
high values observed in utero. Instead of a permanent
postnatal inhibition of insulin secretion we therefore
considered the possibility that an insulin secretagogue
was secreted during late fetal life and removed at birth,
when the cord was cut.

To test this hypothesis, we delivered fetuses one day
before term. The fetuses of half of the litter remained
attached to their placenta and care was taken not to
disturb fetal circulation through the umbilical vessels.
These breathing feto-placental units were transferred to
a warm, humidified and oxigenated incubator. Fetuses of
the other half of the litter were separated from their
placenta and used as controls. As expected,in the controls,
blood glucose concentration fell slowly and plasma IRI
concentration fell rapidly. By contrast, in the feto-
placental units, plasma IRI concentration remained at a
high level and hypoglycemia rapidly ensued (table 1).

Table 1. *Influence of the placenta on plasma IRI and fuels concentration.*

	Control pups, separated from their placenta			Feto-placental units	
Min after delivery	0	30	60	30	60
Blood glucose mg/100 ml	$62.9^{+}_{-}14.3$ (40)	$62.6^{+}_{-}23.9$ (26)	$46.9^{+}_{-}17.7$ (17)	$49.6^{+}_{-}17.3$ (29)	$29.4^{+}_{-}10.7$ (19)
Plasma amino-acids mmol/l	$11.2^{+}_{-}1.3$ (16)	$8.9^{+}_{-}1.3$ (16)	$6.9^{+}_{-}1$ (17)	$11.8^{+}_{-}2.3$ (16)	$10.5^{+}_{-}0.8$ (16)
Plasma IRI µU/ml	$151^{+}_{-}45$ (40)	$73^{+}_{-}28$ (26)	$45^{+}_{-}16$ (17)	$197^{+}_{-}105$ (29)	$109^{+}_{-}49$ (19)

Thus, the sole presence of the placenta maintained hyper-insulinemia but the mechanism whereby it did so remained to be elucidated.

A first possible explanation was that the placenta keeps on supplying aminoacids to the fetus, maintains hyperaminoacidemia and consequently hyperinsulinemia. This hypothesis would also imply that, during fetal life, insulin secretion depends more upon aminoacids than upon glucose since, in our feto-placental units, hyperinsulinemia was maintained in spite of deepening hypoglycemia.

Another possibility is that the placenta triggers insulin secretion by a hormonal mechanism. To test this hypothesis, placentas were homogenized and, after ultra-centrifugation, the supernatant was chromatographed on G50 sephadex. The fractions were I.V. injected to adult rats and their ability to stimulate insulin secretion tested. Secretagogue activity was identified in the first fractions of the gel internal volume, indicating that the

molecular weight of the active substance was greater than
5.000 daltons. The fractions containing the active
substance were rechromatographed on G200 sephadex and the
insulin secretagogue activity recovered in the total
volume of the gel indicating a molecular weight of less
than 30.000 daltons. The activity of this placental
extract is presently investigated on isolated islets but
results are too preliminary to be mentioned. These experiments support the existence of a hormonal factor of
placental origin which stimulates the fetal B cell and
contributes to the maintenance of fetal hyperinsulinemia.

Of course, the potential role of insulin on fetal
metabolism is determined by the rate of insulin secretion
but also by the sensitivity of the target tissues to the
hormone. As the first step of hormone action is binding
to specific recognition sites or receptors, we attempted
to investigate the interaction of insulin with the fetal
tissues in vivo. For this purpose, we used purified carrier-
free ^{125}I-insulin. In this tracer, iodine is mainly
located on the A chain and, inside the A chain, on Tyr^{A14}.
In all the tested systems, this molecule practically had
the same biological potency as unlabeled insulin (18).

Three minutes after injection of monoiodoinsulin into
a vitelline vein of the intact egg in utero, the tracer was
selectively concentrated by 2 organs, the liver and the kidneys. The specific activity of the placenta was low, comparable to that of the carcass. Similar results were obtained
whether the fetuses were tested 1, 3 or 5 days before term.
As previously shown, the concomitant injection of an excess
of unlabeled insulin markedly decreased the percentage
of radioactivity taken up by the liver (saturable mechanism)
while more of the injected radioactivity was recovered
in the kidneys (non saturable mechanism) and in the rest
of the carcass (6). Radioautographs of the liver
demonstrated initial localization of the tracer on the
plasma membranes of the hepatocytes (figure 2). Very few
grains were visible in the vascular spaces. In younger

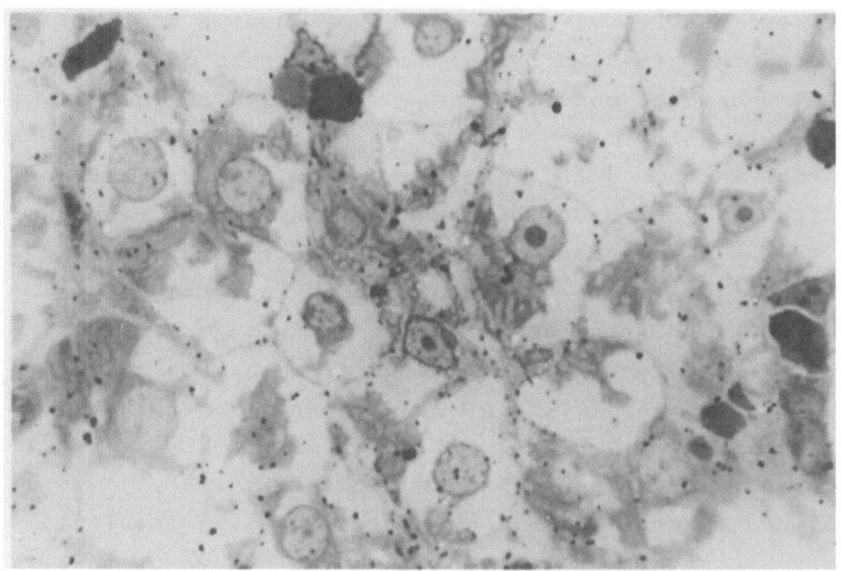

Figure 2: Radioautograph of the liver of a one day preterm fetus
3 min after ^{125}I-insulin injection. The liver was fixed with
glutaraldehyde, postfixed in osmium tetroxide and embedded in Epon.
One μ thick section was coated with Ilford L4 emulsion and developed
after 5 weeks. Note that the silver grains are mainly lined up along
the hepatocytes plasma membranes. Large pale cytoplasmic areas
correspond to glycogen stores (X 330).

fetuses, in which the liver contained a large amount of
hematopoietic tissue, silver grains were associated with
the hepatocytes and to a much smaller extent to the
hematopoietic cells. In the presence of an excess of
insulin, the distribution of the silver grains was
markedly altered : fewer were seen on the hepatocyte
plasma membranes, many were in the vascular compartment.
With regard to the kidneys, silver grains were almost
exclusively concentrated in the proximal tubules.

In the next experiments, the tracer and the excess
unlabeled insulin were injected sequentially in order to
explore the reversibility of the insulin binding mechanisms
in the fetus. Three groups of 1 day preterm fetuses
received the same amount of labeled insulin at time O.
Control rats were bled 3 or 5 min later and, as expected,

plasma monoiodoinsulin concentration rapidly fell.
Experimental pups received an excess of insulin 3 min after
the tracer and were bled 2 min later. In this group, the
amount of undamaged monoiodoinsulin recovered in the plasma
was higher than in the control group at 5 min and even
higher than in the control group at 3 min. Thus, native
insulin not only stopped the escape of the tracer from the
plasma to the tissue compartment but also shifted labeled
insulin from the tissue compartment back into the plasma.
When the delay between the 2 injections increased, less
and less labeled insulin dissociated from the tissue
compartment. The half life of monoiodoinsulin in the
receptor compartment was very short, 6 min. This figure
is comparable to that observed in adult rats (19).

To investigate the ontogeny of the hepatic insulin
receptors, we attempted to quantitate the number of insulin
receptors 5, 3 and 1 day before term. Fetuses were
injected with monoiodoinsulin mixed with increasing amounts
of native insulin and liver radioactivity measured 3 min
later. As shown in figure 3, the liver uptake of labeled
hormone decreased when increasing amounts of native
insulin were coinjected. Liver radioactivity in the
presence of a large excess of native insulin ("non specifically bound radioactivity") was subtracted from each value
of liver radioactivity. Knowing the specific activity of
the injected mixtures and the "specifically bound radioactivity", we calculated the amount of insulin bound to
the liver. In the presence of increasing doses of insulin,
liver bound insulin increased and reached a plateau value
corresponding to saturation of all binding sites. In the
5 day preterm fetuses, liver binding capacity was
evaluated at 1.17 pmol/liver. Under similar conditions,
liver binding capacity was 1.51, 5 and 180 pmol/liver in
3 day preterm, 1 day preterm and adult rats respectively.
Expressed as pmol/g liver, binding capacities were 12.6,
6.7, 15.3 and 15 respectively in 5, 3 and 1 day preterm
fetuses and in adult rats. Although the measurement of

17 day p.c. fetuses

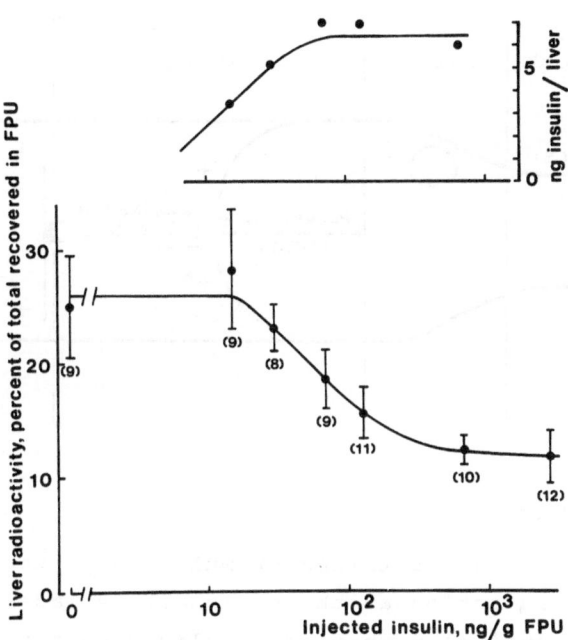

Figure 3: Lower panel: Liver radioactivity 3 min after I V injection of ^{125}I*-insulin alone or mixed with increasing amounts of unlabeled insulin into feto-placental units (FPU). Ave – SD, number of fetuses between parenthesis. Upper panel: plot of the liver bound insulin versus the total injected dose.*

liver binding capacity yields but a crude estimate of true receptor number, our results clearly demonstrate that already 5 days before term, the liver is equipped with an adequate number of insulin receptors.

In figure 4, interactions between fuels, placenta, B cells and insulin target cells are depicted. In fetal blood, aminoacids are higher and glucose lower than in maternal blood, making aminoacids a good and glucose a poor candidate as insulin secretagogue in utero. In addition, a humoral factor of placental origin appears

PLACENTO-INSULAR AXIS : A PROPOSED MODEL

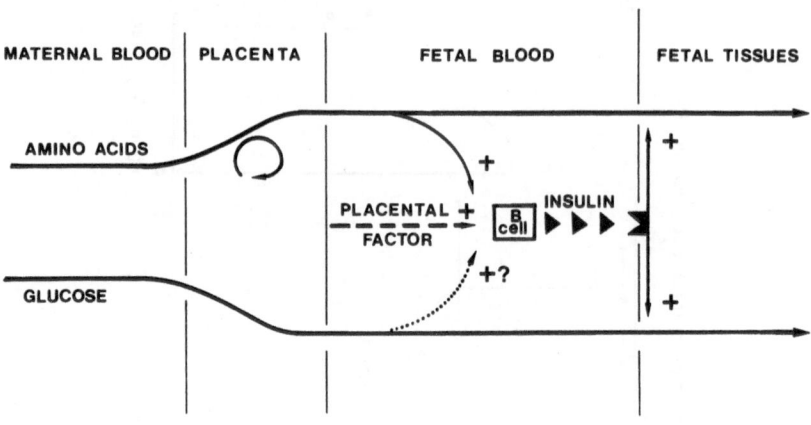

Figure 4: see text.

to contribute to the sustained stimulation of the fetal
B cell.   Insulin action at the periphery is made possible
by an already optimal number of insulin receptors on the
target cells.   At birth, when the cord is cut, the
maternal supply of fuels is interrupted and the just born
pup would be in great danger of developing hypoglycemia
if the factors which maintain fetal hyperinsulinemia
were not, fortunately, removed at the same time.   Later,
when the pups will nurse, fuels and signals triggering
insulin secretion will be provided by the gut, the placento-
insular axis being replaced by the entero-insular one.

ACKNOWLEDGEMENTS

We are grateful to Miss M. Fodor for secretarial
assistance.   This work was supported by grants from the
F.N.R.S., Brussels, Belgium.

REFERENCES

1. Like, AA and LO Orci, Embryogenesis of the human
   pancreatic islets: a light and electron microscopic
   study. Diabetes 21 (suppl. 2): 511-534, 1972.
2. Clark, WR and WJ Rutter, Synthesis and accumulation of
   insulin in the fetal rat pancreas. Develop Biol 29:
   468-481, 1972.
3. Sodoyez-Goffaux, FR, JC Sodoyez, and PP Foà, Effects of
   gestational age, birth and feeding on the insulino-
   genic response to glucose and tolbutamide by fetal
   and newborn rat pancreas. Diabetes 20:586-591, 1971.
4. Felix, JM, MT Sutter, BChJ Sutter, and R Jacquot,
   Circulating insulin and tissular reactivity to insulin in the rat during the perinatal period. Horm
   Metab Res 3:71-75, 1971.
5. Kervran, A and JR Girard, Glucose-induced increase of
   plasma insulin in the rat foetus in utero. J Endocr
   62:545-551, 1974.
6. Sodoyez-Goffaux, FR, JC Sodoyez, and CJ De Vos, Insulin
   secretion and metabolism during the perinatal period
   in the rat. Evidence for a placental role in fetal
   hyperinsulinemia. J Clin Invest 63:1095-1102, 1979.
7. Grasso, SA, A Messina, N Saporito, and G Reitano,
   Serum-insulin response to glucose and aminoacids in
   the premature infant. Lancet II:755-757, 1968.
8. Sodoyez-Goffaux, FR, JC Sodoyez, CJ De Vos, and PP Foà,
   Insulin and glucagon secretion by islets isolated
   from fetal and neonatal rats. Diabetologia 16:121-123,
   1979.
9. Van Assche, FA, W Gepts, and M de Gasparo, The endocrine
   pancreas in anencephalics. Horm Metab Res 1:251-252,
   1969.
10. Martin, JM, CC Mok, J Penfold, NJ Howard, and D Crowne,
    Hypothalamic stimulation of insulin release. J Endocr
    58:681-682, 1973.
11. Beloff-Chain, A, JA Edwardson, and J Hawthorn,
    Corticotrophin-like intermediate lobe peptide as an
    insulin secretagogue. J Endocr 73:28-29P, 1977.
12. Van Assche, FA, Quantitative histology of the pancreas
    in decapitated and normal rat fetuses. Horm Metab Res
    3:285-286, 1971.
13. Girard, JR, A Kevran, and R Assan, Functional maturation
    of the A cell in the rat, In : Early Diabetes in
    Early Life, Camerini-Davalos, RA and HS Cole (eds),
    Academic Press Inc, New York, 57-71, 1975.
14. Bataille, D, C Jarrousse, N Vauclin, C Gespach, and
    G Rosselin, Effect of vasoactive intestinal peptide
    (VIP) and gastric inhibitory peptide (GIP) on
    insulin and glucagon release by perifused newborn rat
    pancreas, In: Glucagon : Its Role in Physiology and
    Clinical Medicine, Foà PP, JS Bajaj, and N Foà (eds),
    Springer-Verlag New York, 255-269, 1977.

15. Kervran, A, J Randon, and JR Girard, Dynamics of glucose-induced plasma insulin increase in the rat fetus at different stages of gestation. Effects of maternal hypothermia and fetal decapitation. Biol Neonate 35:242-248, 1979.

16. Blouquit, MF, G Sturbois, G Breart, C Grill, C Sureau, and J Roffi, Catecholamine levels in newborn human plasma in normal and abnormal conditions and in maternal plasma at delivery. Experientia 35:618-619, 1979.

17. Grajwer, LA, MA Sperling, J Sack, and DA Fisher, Possible mechanisms and significance of the neonatal surge in glucagon secretion: studies in newborn lambs. Pediat Res 11:833-836, 1977.

18. Sodoyez, JC, F Sodoyez-Goffaux, MM Goff, AE Zimmerman, and ER Arquilla, [127]I- or carrier-free ^{125}I-monoiodo-insulin. Preparation, physical, immunological,and biological properties, and susceptibility to "insulinase" degradation. J Biol Chem 250:4268-4277, 1975.

19. Sodoyez, JC, F Sodoyez-Goffaux and YM Thiry-Moris. ^{125}I-Insulin. Kinetics of interaction with its receptors and rate of degradation in vivo. Am J Physiol. In press.

DISCUSSION

Qu.Jones : To what extent does sympathetic control of insulin secretion account for the inhibition of insulin secretion at birth?

An.Sodoyez : Sympathetic activity is very likely increased
at birth, particularly when the newborn is anoxic or left
in a cold environment. This increased sympathetic activity
may contribute to the postnatal inhibition of insulin secretion but we do not think that it is the only factor which
modulates B cell function after cord cutting since increased
sympathetic activity is but a transient phenomenon whereas
the postnatal decrease of plasma IRI will last permanently.

Com.Jones: ACTH-related peptides are secreted from the placenta and may represent a stimulatory substance for insulin
secretion from the fetal pancreas.

An.Sodoyez : At the present time, we have too little information on the chemical structure of the insulin secretagogue
placental factor. It could be either a unique substance
synthetized by the placenta or a peptide related to hormones
normally produced by the pituitary,the gastrointestinal
tract, etc... Among the many possibilities, it could be an
ACTH related peptide.

Com.Teller : It has been known for some time that neonatal
rats fail to increase their insulin release upon stimulation with increasing concentrations of glucose. The question arose, whether this phenomenon is due to impaired insulin biosynthesis or secretion?
Heinze et al. (Ped.Res. 9:670, 1975; Diabetes 24:373,1975)
studied isolated islets of 21-day-old fetal rats. The in -
sulin biosynthesis could be increased only by increasing
glucose concentrations up to 100 mg/dl. The conversion of
proinsulin to insulin was greater in fetal islets than in
those of newborn 5 or 10-day-old rats. It could be augmented further by 24-36 hour infusion of glucose to the mothers
prior to delivery. From these data it appears, that β -
cells of fetal rats may be able to adjust their insulin biosynthesis by various concentrations of glucose. This is

achieved mainly by adjusting the conversion of proinsulin
to insulin.

The stimulation of <u>insulin secretion</u> by glucose was impaired
in newborn rats up until the 5th day of life.

The offsprings of mothers made hyperglycemic for a week,
however, responded with increased insulin secretion upon
challenges by 100 mg and 150 mg of glucose per dl.

Some similarities exist in the dynamics of insulin secretion and biosynthesis during the perinatal period and the
remission phase of juvenile-onset diabetes. In both situations arginine appears to be the most effective secretory
agent ,while stimulation by glucose alone seems to be impaired.

GLUCAGON SECRETION DURING THE PERINATAL PERIOD

Jean Girard and Roger Assan

INTRODUCTION

The ontogenetic differentiation of the endocrine pancreas has been reviewed recently (1). It appears that A cells and glucagon are present in the pancreas early in development, a long time before the specific target organs of this hormone would be differentiated. It has been postulated that glucagon might play a role in regulating growth and differentiation of embryonic cells (2), but it has not been demonstrated that glucagon was secreted at these early stages. During late fetal life, exogenous glucagon produced several metabolic effects in the fetus. Infusion of glucagon in fetal lamb during the last third of gestation results in significant increases in their plasma glucose concentration (3, 4). The effect is probably due entirely to stimulation of fetal hepatic glycogenolysis since no significant gluconeogenesis can be demonstrated in these conditions (4). Glucagon injection in term rat (5) and monkey (6) foetuses also results in an increase of blood glucose. In the rat, injection of glucagon produces a marked decrease in fetal liver glycogen concentration (5) and the premature appearance of phosphoenolpyruvate carboxykinase (PEPCK), the rate limiting enzyme of liver gluconeogenesis (5, 7, 8). Furthermore, injection of glucagon to the rat fetus, induces a decrease in fetal plasma amino acids and stimulates amino acid uptake by fetal liver (9). Glucagon has also been reported to stimulate glycogenolysis (10) and to induce PEPCK (11) in cultured fetal rat hepatocytes. As exogenous glucagon was capable to reproduce in utero the metabolic changes in glucose and amino acid metabolism

which normally occur immediately after birth i.e. liver
glycogenolysis, induction of hepatic PEPCK and appearance
of gluconeogenesis (see 12 for a review), it has been postulated that this hormone might play an important role during this critical period of development (13).
The aim of the present paper is to review the different
factors which might affect the secretion of glucagon by the
pancreas during late, fetal and neonatal periods.

METHODOLOGICAL ASPECTS RELATED TO PLASMA GLUCAGON
DETERMINATION

Determination of glucagon in plasma by radioimmunoassay can
be fraught with interferences due to a glucagon-like material of enteric origin (gut glucagon-like immunoreactivity)
and a plasma globulin commonly denominated big plasma glucagon (BPG). Both of these interfering substances are devoid of the biological activity of true pancreatic glucagon (molecular weight 3.500 daltons). Most radioimmunoas-
says included C-terminal antisera admittedly specific for
pancreatic glucagon since they do not cross react with
gut glucagon-like substances. Big plasma glucagon (BPG) in-
terferes with these "specific" antisera but its concentration does not change during experiments of short duration.
Thus BPG can contribute to overestimate the basal levels
of plasma glucagon in unextracted plasma but not to the
changes of plasma glucagon during acute spontaneous or experimental conditions (14). In this paper, the term plasma
glucagon will be used to refer to plasma immunoreactive
glucagon determined by using only antisera "specific" for
true pancreatic glucagon.

GLUCAGON SECRETION DURING FETAL LIFE

The presence of glucagon at concentrations in the same
range as those in maternal blood has been found in the
blood of near term fetuses from several species: Rat (15-
17), Rabbit (18), Sheep (19-21), Monkey (6) and in umbilical blood of human at delivery (17, 22-25). As glucagon

does not cross the placenta in the Rat (15), Sheep (18, 26 27), Monkey (6) and Human (28, 29), glucagon circulating in fetal plasma is released by the fetal pancreas. In the rat, there is no clear relationships between the level of circulating glucose, total amino nitrogen and free fatty acids (F.F.A.) and the concentration of glucagon in fetal plasma during the last 4 days of gestation (30). At present, the factor(s) which control the secretion of glucagon by the fetal pancreas in basal conditions (i.e. in the fetus of fed mothers) are unknown.

1. <u>Effects of hypo or hyperglycemia</u>

Some attempts have been made to study plasma glucagon in experimental conditions in which fetal blood glucose was acutely modified by perfusion of glucose or insulin to the mother (16).

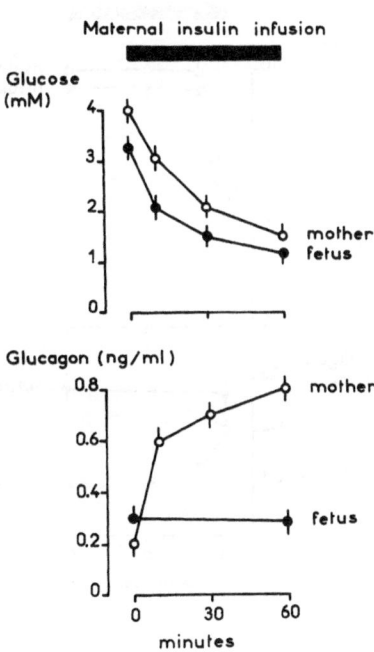

Figure 1 : Effects of insulin-induced hypoglycemia on plasma glucagon levels in term pregnant rats and their fetuses.

As it could be expected, hyperglycemia produces a decrease
in maternal plasma glucagon and insulin-induced hypoglycemia stimulates maternal glucagon secretion (Figures 1 & 2).
By contrast, no change in fetal plasma glucagon occurs during fetal hyperglycemia or hypoglycemia (Figures 1 & 2).
Similarly, plasma glucagon concentrations in the fetal lamb
chronically catheterized in utero (20, H.G. Shelley and J.
R. Girard, unpublished observations) or exteriorized under
anesthesia during acute studies (31) are not significantly
influenced by hyper or hypoglycemia. The unresponsiveness
of fetal pancreatic A cells to changes in fetal blood glucose concentrations has also been documented in the monkey
(6). The secretion of glucagon in vitro by pieces of pancreas or isolated islets of rat (32, 33), mouse (34) and

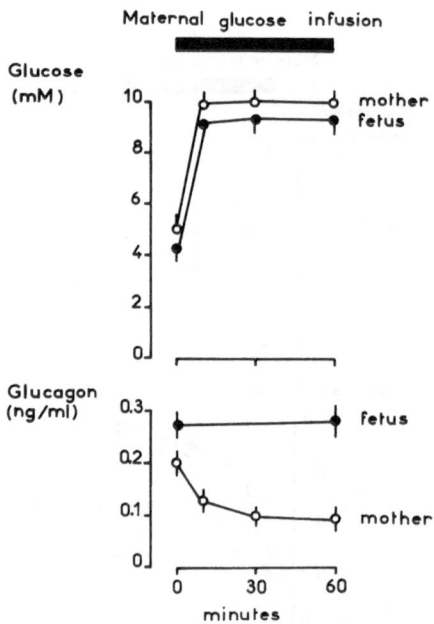

Figure 2 : Effect of hyperglycemia on plasma
glucagon levels in the term pregnant rats and
their fetuses.

human (35) foetuses, is uninfluenced by large variations
in medium glucose concentrations. All these observations
clearly show that acute changes in blood glucose concentrations in the fetus are relatively unimportant in influencing glucagon secretion by the fetal pancreas.
By contrast several lines of evidence suggest that chronic
changes in blood glucose concentration in the fetus affect
the secretion of glucagon by the fetal pancreas. In pregnant rats chronically catheterized, the continuous infusion of glucose during the last 3 days of gestation in
order to raise the maternal and fetal blood glucose levels
in the range of 10 mmoles/l, induces a 2 fold decrease in
fetal plasma glucagon at term (A. Ktorza, L.Picon, B.
Portha, and J.Girard, unpublished observations). Moreover,
fasting of pregnant rats during the last 4 days of gestation, is associated with a chronic fetal hypoglycemia and
with a 2 fold increase in fetal plasma glucagon (36). These observations suggest that chronic exposure to high or
low glucose concentrations can affect the secretion of
the fetal rat pancreas.

2. Effects of amino acids
There are some discrepancies about the sensitivity in vivo
of fetal A cells to amino-acids. In vivo, infusion of
alanine or glycine in chronically catheterized fetal lambs
failed to cause significant stimulation of glucagon secretion (20, 21). Similarly, alanine was ineffective in stimulating glucagon release in the fetal monkey (6). However,
arginine infusion into exteriorized fetal lambs (31) or
in the rat fetus in utero (16) caused a significant increase in glucagon secretion. In humans at term, the infusion
of alanine into the mother during labor caused a significant increase in umbilical cord plasma glucagon (25), suggesting that term human fetal pancreas was sensitive to
amino-acids.

By contrast, in vitro studies with fetal pancreas pieces
or isolated islets of lamb (37), rat (18, 33), mouse (34)
or human (18, 35) have shown that glycine, alanine and
arginine markedly stimulated glucagon release.
The reason why fetal pancreatic A cells are sensitive to
amino acids in vitro and much less, or not at all, in vivo
could be due to the already very high level of circulating
amino acids in fetal blood thus blunting the effect of
infused amino acids.

3. Effects of neurotransmitters

In adult animals, plasma glucagon increases markedly in
response to adrenergic stimulation (review in 38). Norepinephrine and epinephrine injections in fetal rats increases plasma glucagon (16) and intravenous infusion of
epinephrine into chronically catheterized fetal lamb significantly stimulates glucagon secretion (21, 39). Propranolol markedly attenuated the rise in glucagon secretion during epinephrine infusion in fetal lamb (40) suggesting
that stimulation of β-adrenergic receptors are involved
in the effect of epinephrine. L-Dopa injected intravenously
to the monkey fetus increases fetal plasma glucagon within
20 minutes (40). In the rat fetus, serotonine and dopamine
injections remain ineffective on fetal plasma glucagon
(16, 30). Exogenous acetylcholine also increases plasma
glucagon in the rat fetus although the rise is of smaller
magnitude than that observed after catecholamine injection
(16).
Similar effects of epinephrine and norepinephrine on glucagon secretion have also been observed in vitro in fetal
pancreas pieces of rat (18), human (18) and lamb (21) fetuses.

4. Effects of hypoxia

Hypoxia induced by ligation of uterine arteries results in
a marked rise in fetal plasma glucagon (Figure 3).

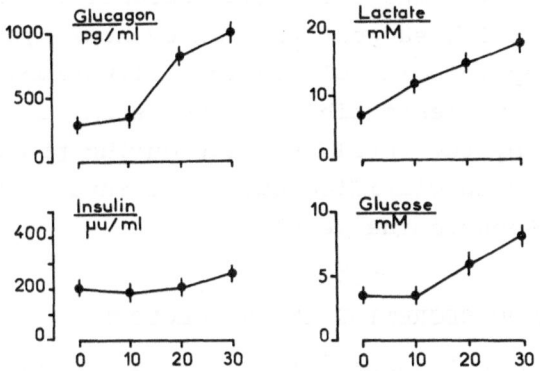

Minutes after clamping of uterine arteries

Figure 3 : Effects of hypoxia induced by
clamping of uterine arteries on plasma
insulin and glucagon and on blood glucose
and lactate levels in the rat fetuses.

In chronically catheterized fetal lamb, plasma glucagon
is increased in hypoxic fetuses when the pH falls below
7.30 (H.J. Shelley and J. Girard, unpublished observations). The glucagon levels in umbilical blood of infant
with fetal distress (pH $<$ 7.2 and hypoxic) are 2 fold higher than in normal infants (23). The mechanisms by which
hypoxia results in increased glucagon release have not
been clarified, but could be due to an activation of sympathetic nervous system. Indeed, hyperglucagonemia during
acute hypoxia in the dog is largely mediated by adrenergic mechanisms involving α-adrenergic receptor activation
(41).

5. Prolonged pregnancy

Prolongation of pregnancy of 2 days in the rat, by subcutaneous progesterone injection, is associated with a marked rise in plasma glucagon and a fall in plasma insulin (13, 42). The postmature fetuses show a large decrease in liver glycogen content (43) and a 10 fold increase in liver PEPCK (44, 45), which suggest that endogenous hyperglucagonemia plays a physiologic role in this situation. The factors which could be responsible for the stimulation of glucagon secretion by the fetal pancreas during prolonged gestation have not been clarified but could involve hypoxia or increased catecholamine levels.

GLUCAGON SECRETION IN THE NEWBORN

At birth the maternal supply of nutrients to the fetus ceases abruptly and the circulating substrates change rapidly. There is a marked fall in plasma amino-acid concentrations and a transient decline of blood glucose concentration in rat (46) and human newborns (47). By contrast, in the newborn sheep , blood glucose concentration, which was already very low in utero, shows very little change (48). During the immediate postnatal period a significant increase in plasma glucagon occurs in all the species studied. This was first described in the rat (49) and then confirmed in other species : human (47), sheep (48) and more recently in rabbit (50). In the rat, the rise in plasma glucagon is observed both after caesarian section (42, 46, 49, 51) or vaginal delivery (13, 17), and it occurs similarly both in fasted or suckled newborn rats (52).

1. Evidence for a role of sympathetic nervous system in the neonatal surge of glucagon

It has been postulated by several authors that the transient postnatal hypoglycemia which occurs in the rat maintained at thermoneutrality (37°C), might trigger the release of glucagon by the pancreas. This is not the case since similar changes occur in the newborn maintained at

24°C where blood glucose levels do not fall (53). Furthermore the surge of glucagon is also observed in species
(sheep, rabbit) in which there is no fall of blood glucose
at birth (48, 50). Finally, the insensitivity of newborn
 A cells to changes in glucose concentration is well documented both <u>in vivo</u> (9, 54, 55) and <u>in vitro</u> (17, 32-34,
56). Although amino acids have been reported to be effective stimulators of glucagon release by the newborn pancreas (56-58), it is unlikely that they play a role in
the surge of glucagon at birth, since plasma amino acids
fall dramatically in the immediate postnatal period (46).
A more likely mechanism to explain the acute rise in plasma glucagon at birth might relate to an adrenergic stimulation in response to the stress of birth (transient hypoxia, cold exposure, or cord cutting , 48). Epinephrine
and norepinephrine levels are increased several fold in
umbilical blood of human newborns (59, 60) and in newborn
rats (61) and both of these hormones are capable of stimulating glucagon release by the fetal pancreas (16, 57).
A direct stimulation of neonatal pancreas through sympathetic nerves is also possible since it has been shown
that nervous structures are present adjacent to endocrine
islets in the rat fetus (62). As α-adrenergic blocking
drug (phentolamine) partially inhibits the rise in plasma
glucagon in the newborn rat, while the β-adrenergic blocking drug (propranolol) do not (30), it is suggested that
an α-adrenergic mechanism is involved in the neonatal
surge of glucagon.

2. Glucagon secretion in the newborn infants to diabetic
   mothers (IDM)

Infants born to diabetic mothers have an increased incidence of hypoglycemia in the immediate postnatal period. For
a long time, this has been considered as the result of
their hyperinsulinemia and of their increased glucose utilization. More recently it has been reported that IDM show
also a decreased glucose production during the perinatal
period (63).
The spontaneous increase in plasma glucagon which occurs
in normal human newborn infants (22, 23, 47) is blunted in
IDM (22, 64). This failure of glucagon release could contribute in association with hyperinsulinemia to the defect
of hepatic glucose production in IDM. At present we do not
know the factors responsible for this failure, but it may
be related to deficient catecholamine release at birth
(65, 66) although a recent (67) report do not confirm these previous observations.

CONCLUSIONS

The essential role of glucagon in glucose homeostasis in
the newborn has been recently demonstrated by Sperling and
his collaborators. Infusion of somatostatin in newborn
lambs aged 24-72 hours suppressed both insulin and glucagon and induced an hypoglycemia (68). When glucagon was infused with somatostatin, producing an insulin deficiency
concomitantly with high plasma glucagon, the blood glucose
level was restored to normal levels (68). Moreover it has
been reported that human newborns suffering of glucagon
deficiency develop a profound hypoglycemia (69-71), which
is rapidly corrected by exogenous glucagon administration
(69-71).
All these data strongly support the view that glucagon
plays an important role in neonatal metabolic adaptations.

REFERENCES

1. Pictet, R, and WJ Rutter, Development of the embryonic endocrine pancreas, In : <u>Endocrine Pancreas,</u> Freinkel N, and DF Steiner (eds), Baltimore, Williams and Wilkins Inc, 25-66, 1972

2. Rall, LB, RL Pictet, RH Williams, and WJ Rutter, Early differentiation of glucagon-producing cells in embryonic pancreas : a possible developmental role of glucagon. Proc Nat Acad Sci, USA 70 : 3478-3482,1973

3. Bassett, JM, and GD Thornburn, The regulation of insulin secretion by the ovine fetus <u>in utero</u>. J Endocr 62 : 59-74, 1971

4   Warnes, DM, RF Seamark, and FJ Ballard, The appearance of gluconeogenesis at birth in sheep. Activations of the pathway associated with blood oxygenation. Biochem J 162 : 617-626, 1977

5. Girard, JR, D Caquet, D Bal, and I Guillet, Control of rat liver phosphorylase, glucose-6-phosphatase and phosphoenolpyruvate carboxykinase activities by insulin and glucagon during the perinatal period. Enzyme 15 : 272-285, 1973

6. Chez, RA, DH Mintz, MF Epstein, AR Fleischman, GK Oakes, and DL Hutchinson, Glucagon metabolism in nonhuman primate pregnancy. Amer J Obstet Gynecol 120 : 690-694, 1974

7. Yeung, D, and IT Oliver, Factors affecting the premature induction of phosphopyruvate carboxylase in neonatal rat liver. Biochem J 108 : 325-331, 1968

8. Hanson, RW, L Fisher, FJ Ballard, and L Reshef, The regulation of phosphoenolpyruvate carboxykinase in fetal rat liver. Enzyme 15 : 97-110, 1973

9   Girard, JR, I Guillet, J. Marty, R Assan, and EB Marliss, Effects of exogenous hormones and glucose on plasma levels and hepatic metabolism of amino-acids in the fetus and in the newborn rat. Diabetologia 12 : 327-337, 1976

10 Plas, C, and J. Nunez, Glycogenolytic response to glucagon of cultured fetal hepatocytes. Refractoriness following prior exposure to glucagon. J Biol Chem 250 : 5304-5311, 1975

11. Bulanyi, GS, JG Steele, MC McGrath, GCT Yeoh, and IT Oliver, Hormonal regulation of phosphoenolpyruvate carboxykinase in cultured fetal hepatocytes from rat. Europ J Biochem 102 : 93-100, 1979

12. Ballard, FJ, Gluconeogenesis and the regulation of blood glucose in the neonate, In : Diabetes, Rodriguez, RR, and J Vallance-Owen (eds), Amsterdam, Excerpta Medica, 592-600, 1971

13. Girard, JR, P Ferré, A Kervran, JP Pégorier, and R Assan, Role of the insulin/glucagon ratio in the changes of hepatic metabolism during development of the rat, In : Glucagon : its Role in Physiology and Clinical Medicine, Foa, PP, JS Bajaj, and NL Foa (eds), New York, Springer Verlag, 563-581, 1977

14. Weir, GC, Assessment of glucagon immunoreactivity in plasma, In : Glucagon : its Role in Physiology and Clinical Medicine, Foa,PP, JS Bajaj, and NL Foa (eds), New York, Springer Verlag, 65-76, 1977

15. Girard, J, R Assan, and A Jost, Glucagon in the rat foetus, In : Foetal and Neonatal Physiology, Comline KS, KW Cross, GS Dawes, and PW Nathanielsz (eds), Cambridge, Cambridge University Press, 456-461, 1973

16. Girard, JR, A Kervran, E Soufflet, and R Assan, Factors affecting the secretion of insulin and glucagon by the rat fetuses. Diabetes 23 : 310-317, 1974

17. Blazquez, E, T Sugase, M Blazquez, and PP Foa, Neonatal changes in the concentration of rat liver cyclic AMP and of serum glucose, FFA, insulin, pancreatic and total glucagon in man and in the rat. J Lab Clin Med 83 : 957-967, 1974

18. Assan, R, JR Attali, G Ballerio, JR Girard, M Hautecouverture, A Kervran, PF Plouin, G Slama, E Soufflet, G Tchobroutsky, and A Tiengo, Some aspects of the physiology of glucagon, In : Diabetes, Malaisse, WJ, and J Pirart (eds), Amsterdam, Excerpta Medica, 144-179, 1974

19. Alexander, DP, R Assan, HG Britton, and DA Nixon, Glucagon in the fetal sheep. J Endocr 51 : 597-598,1971

20. Fiser, RH jr, A Erenberg, MA Sperling, WH Oh, and DA Fisher, Insulin-glucagon substrate interrelations in the fetal sheep. Pediatr Res 8 : 951-955, 1974

21. Bassett, JM, Glucagon, insulin and glucose homeostasis in the fetal lamb. Ann Rech Vet 8 : 362-373, 1977

22. Bloom, SR, and DI Johnston, Failure of glucagon relea-
    se in infants of diabetic mothers. Brit Med J 4 :
    453-454, 1972

23. Johnston, DI, and SR Bloom, Plasma glucagon levels in
    the term human infant and effect of hypoxia. Arch
    Dis Child 48 : 451-454, 1973

24. Milner, RDG, SK Chouksey, KNP Mickleson, and R Assan,
    Plasma pancreatic glucagon and insulin:glucagon
    ratio at birth. Arch Dis Child 48 : 241-242, 1973

25. Wise, JK, SS Lyall, R Hendler, and P Felig, Evidence
    of stimulation of glucagon secretion by alanine in
    the human fetus at term. J Clin Endocr Metab 37 :
    345-348, 1973

26. Alexander, DP, R Assan, HG Britton, and DA Nixon, Im-
    permeability of the sheep placenta to glucagon.
    Biol Neonate 23 : 391-402, 1973

27. Sperling, MA, A Erenberg, RH Fiser, W Oh, and DA
    Fisher, Placental transfer of glucagon in sheep.
    Endocrinol 93 : 1435-1438, 1973

28. Adam, PAJ, KC King, R Schwartz, and K Teramo, Human
    placental barrier to ^{125}I-glucagon early in gesta-
    tion. J Clin Endocrinol 34 : 772-785, 1972

29. Johnston, DI, SR Bloom, KR Greene, and RW Beard, Fai-
    lure of the human placenta to transfer pancreatic
    glucagon. Biol Neonate 21 : 375-380, 1972

30. Girard, JR, A Kervran, and R Assan, Functional matura-
    tion of the A cell in the Rat, In : Early Diabetes
    in Early Life, Camerini-Davalos, RA, and HS Cole
    (eds), New York, Academic Press, 57-71, 1975

31. Alexander, DP, R Assan, HG Britton, E Fenton, and D
    Redstone, Glucagon release in the sheep fetus. 1)
    Effect of hypo and hyperglycemia and arginine. Biol
    Neonate 30 : 1-10, 1976

32. Bajaj, JS, and KD Buchanan, Glucose homesotasis in the
    newborn, In : Glucagon : its Role in Physiology and
    Clinical Medicine, Foan PP, JS Bajaj, and NL Foa
    (eds), New York, Springer Verlag, 583-593, 1977

33. Sodoyez-Goffaux, F, JC Sodoyez, CJ De Vos, and PP Foa,
    Insulin and glucagon secretion by islets isolated
    from fetal and neonatal rats. Diabetologia 16 : 121-
    123, 1979

34. Lernmark, A, and BI Wenngren, Insulin and glucagon release from the isolated pancreas of foetal and newborn mice. J Embryol Exper Morph 28 : 607-614, 1972

35. Schaeffer, LD, ML Wilder, and RH Williams, Secretion and content of insulin and glucagon in human fetal pancreas slices in vitro. Proc Soc Exper Biol Med 143 : 314-319, 1973

36. Girard, JR, P Ferré, M Gilbert, A Kervran, R Assan, and EB Marliss, Fetal metabolic response to maternal fasting in the rat. Amer J Physiol 232 : E456-E463, 1977

37. Bassett, JM, V Hunziker, and D Madill, Glycine and alanine regulation of glucagon secretion in foetal and post-natal lambs. J Physiol (London) 275 : 51-52P, 1977

38. Woods, SC, and D Porte jr, Neural control of the endocrine pancreas. Physiol Rev 54 : 596-619, 1974

39. Christensen, RA, MA Sperling, S Ganguli, and R Anand, Adrenergic modulation of pancreatic hormone secretion in utero : studies in fetal sheep. Pediatr Res in press, 1980

40. Epstein, M, RA Chez, GK Oakes, and DH Mintz, Fetal pancreatic glucagon responses in glucose-intolerant nonhuman primate pregnancy. Amer J Obstet Gynecol 127 : 268-272, 1977

41. Baum, D, D Porte jr, and J Ensinck, Hyperglucagonemia and α-adrenergic receptors in acute hypoxia. Amer J Physiol 237 : E404-E408, 1979

42. Portha, B, L Picon, and G Rosselin, Postmaturity in the rat : high levels of glucagon in the plasma of the foetus and the neonate. J Endocr 77 : 153-154, 1978

43. Portha, B, G Rosselin, and L Picon, Postmaturity in the rat : impairement of insulin, glucagon and glycogen stores. Diabetologia 12 : 429-436, 1976

44. Pearce, PH, BJ Buirchell, PK Weaver, and IT Oliver, The development of phosphopyruvate carboxylase and gluconeogenesis in neonatal rats. Biol Neonate 24 : 320-329, 1974

45. Portha, B, E Le Prevost, L Picon, and G Rosselin, Post-maturity in the rat : phosphorylase, glucose-6-phosphatase and phosphoenolpyruvate carboxykinase activities in the fetal liver. Horm Metab Res 10 : 141-144, 1978

46. Girard, JR, GS Cuendet, EB Marliss, A Kervran, M Rieutort, and R Assan, Fuels, hormones and liver metabolism at term and during the early postnatal period in the rat. J Clin Invest 52 : 3190-3200, 1973

47. Sperling, MA, PV De Lamater, D Phelps, RH Fiser, W Oh, and DA Fisher, Spontaneous and amino acid stimulated glucagon secretion in the immediate postnatal period. Relation to glucose and insulin. J Clin Invest 53 : 1159-1166, 1974

48. Grajwer, LA, MA Sperling, J Sack, and DA Fisher, Possible mechanisms and significance of the neonatal surge in glucagon secretion : studies in newborn lambs. Pediatr Res 11 : 833-836, 1977

49. Girard, JR, D Bal, and R Assan, Glucagon secretion during the early postnatal period in the rat. Horm Metab Res 4 : 168-170, 1972

50. Callikan, S, P Ferré, JP Pégorier, EB Marliss, R Assan, and JR Girard, Fuel metabolism in fasted newborn rabbits. J Develop Physiol 1 : 267-281, 1979

51. Di Marco, PN, AV Ghisalberti, CE Martin, and IT Oliver, Perinatal changes in liver corticosterone, serum insulin and plasma glucagon and corticosterone in the rat. Eur J Biochem 87 : 243-247, 1978

52. Girard, JR, P Ferré, JP Pégorier, A Leturque, and S Callikan, Factors involved in the development of hypoglycemia in fasting newborn rats, In : 2nd European Symposium on Hypoglycemia, Andreani, D, P Lefebvre, and V Marks (eds), New York, Academic Press, in press, 1980

53. Kervran, A., M Gilbert, JR Girard, R Assan, and A Jost, Effect of environmental temperature on glucose-induced insulin response in the newborn rat. Diabetes 25 : 1026-1030, 1976

54. Fiser, RH jr, DL Phelps, PR Williams, MA Sperling, DA Fisher, and W Oh, Insulin-glucagon substrate inter-relationships in the neonatal sheep. Amer J Obstet Gynecol 120 : 944-950, 1974

55. Luyckx, A, F Massi-Benedetti, A Falorni, and PJ Lefeb-
    vre, Presence of pancreatic glucagon in the portal
    plasma of human neonates. Differences in the insulin
    and glucagon responses to glucose between normal in-
    fants and infants from diabetic mothers. Diabetolo-
    gia 8 : 296-300, 1972

56. Edwards, JC, K Asplund, and G Lundquist, Glucagon re-
    lease from the pancreas of the newborn rat. J
    Endocr 54 : 493-504, 1972

57. Marliss, EB, CB Wollheim, B Blondel, L Orci, AE Lambert,
    W Stauffacher, AA Like, and AE Renold, Insulin and
    glucagon release from monolayer cell cultures of
    pancreas from newborn rat. Europ J Clin Invest 3 :
    16-26, 1973

58. Jarousse, C, and G Rosselin, Interaction of amino acids
    and cyclic AMP on the release of insulin and gluca-
    gon by newborn rat pancreas. Endocrinol 96 : 168-177
    1975

59. Lagercrantz, H, and P Biotoletti, Catecholamine release
    in the newborn infant at birth. Pediatr Res 11 : 889-
    893, 1977

60. Eliot, RJ, R Lam, RD Leake, CJ Hobel, and DA Fisher,
    Plasma catecholamine concentrations in infants at
    birth and during the first 48 hours of life. J Pe-
    diatr in press, 1980

61. Kervran, A, J Randon, and JP Maltier, Effets de la tem-
    pérature ambiante sur les sécrétions hormonales pan-
    créatiques à la naissance, In : Journées de Diabéto-
    logie de l'Hôtel-Dieu , Paris, Flammarion, 165-174,
    1979

62. Perrier, H, Evolution de l'ultrastructure du pancréas
    chez le foetus de rat. Diabetologia 6 : 605-615,
    1970

63. Kalhan, SC, SM Savin, and PAJ Adam, Attenuated glucose
    production rate in newborn infant of insulin depen-
    dent diabetic mothers. New England J Med 296 : 375-
    376, 1977

64. Williams, PR, MA Sperling, and Z Racasa, Blunting of
    spontaneous and alanine-stimulated glucagon secre-
    tion in newborn infants of diabetic mothers. Amer J
    Obstet Gynecol 133 : 51-56, 1979

65. Light, IJ, JM Sutherland, JM Loggie, and TE Gaffney, Impaired epinephrine release in hypoglycemic infants of diabetic mothers. New England J Med 277 : 394-398 1967

66. Stern, L, A Ramos, and J Leduc, Urinary catecholamines excretion in infants of diabetic mothers. Pediatrics 42 : 598-605, 1968

67. Young, JB, WR Cohen, EB Rappaport, and L Lansberg, High plasma norepinephrine concentrations at birth in infants of diabetic mothers. Diabetes 28 : 697-699, 1979

68. Sperling, MA, L Grajwer, RD Leake, and DA Fisher, Effects of somatostatin (SRIF) infusion on glucose homeostasis in newborn lambs : evidence for a significant role of glucagon. Pediatr Res 11 : 962-967, 1977

69. Sherwood, WG, GW Chance, CJ Toews, JM Martin, and EB Marliss, A new syndrome of familial pancreatic agenesis : essential role of glucagon in neonatal gluconeogenesis. Pediatr Res 8 : 438 abst, 1974

70. Vidnes, J, and S Oyasaeter, Glucagon deficiency causing severe neonatal hypoglycemia in a patient with normal insulin secretion. Pediatr Res 11 : 943-949,1977

71. Kollee, LA, LA Monnens, V Cejka, and RH Wilms, Persistent neonatal hypoglycemia due to glucagon deficiency. Arch Dis Child 53 : 422-424, 1978

DISCUSSION

Qu.Moses : How do you explain the low insulin and high glucagon secretion rates found in fasted, phloridzine treated
or prolonged gestational animals in terms of neurotransmit-
ter effect on the alpha cells of the pancreas in the fetal
or neonatal rat?

An.Girard : An increased activity of sympathetic nervous
system could be expected, secondary to hypoxia, in fetuses
during prolonged gestation and could explain the low insulin and high glucagon levels observed in fetal plasma.
In fetuses from fasted mothers or from phloridzin injected
mothers, chronic fetal hypoglycemia could also be an explanation.

Qu.Teller : You did not mention the specificity of your
assay. What "type" of glucagon have you been measuring?
To my knowledge the last answer to the problem of specificity of glucagon assay has not yet been given (See Maier,
Horm.Metab.Res.1978).

An.Girard : We have measured glucagon like immunoreactivity
by using an C-terminal antisera (30K) provided by R.H. Unger (Dallas, Texas) admittedly specific of true pancreatic
glucagon. Although it has been shown that this antiserum
also cross reacts with plasma globin but, as this interference factor does not change in the same individual during
acute conditions, the observed variations induced by various
treatments can be considered as the reflect of changes in
true pancreatic glucagon.

Qu.Bossi : The immediate surge of glucagon after cord clamping, 30 minutes after birth, resembles very much the immediate rise in thyroxine levels after cord clamping at the
same postnatal age shown by Fisher's group in Los Angeles.
Is there anything known about postnatal glucagon rise in
hypothyrotic animals or human newborns?

An.Girard : No.

Com.Britton : We have measured glucagon in the arterial

blood of exteriorized sheep foetuses and followed its level
after tying the cord and allowing the lamb to breathe.
Our values for pancreatic glucagon (unpublished) are similar to those reported by Fisher et al. but we do not see a
peak at 60 minutes.

PLASMA INSULIN AND GLUCAGON IN WELL-OXYGENATED AND HYPOXIC FETAL LAMBS

Heather J. Shelley and J.R. Girard

INTRODUCTION

It was believed until recently that the fetus had little need to control its environment, since this would be done indirectly via the mother's homeostatic mechanisms. The close relationship between maternal and fetal plasma glucose and the instability of plasma glucose levels in the newborn suggested that the ability to regulate glucose uptake and metabolism was not developed until after birth(1,2). The fetus was thought to receive a constant supply of glucose from its mother via the placenta. But the trans-placental passage of glucose varies with both the maternal plasma glucose and the maternal placental blood flow (3). When the latter is compromised, the fetus becomes hypoxic and the ability to regulate its glucose metabolism could be of vital importance for fetal survival.

In the adult, minute by minute regulation of glucose uptake and metabolism is achieved by changes in the secretion of the pancreatic hormones insulin and glucagon, insulin promoting glucose uptake and metabolism when the supply is plentiful and glucagon stimulating hepatic glycogenolysis and gluconeogenesis when glucose production is needed (4). The present work compares the plasma levels of these hormones in well-oxygenated and hypoxic fetal lambs exposed to changes in the glucose supply. Indwelling carotid artery and jugular vein catheters were inserted in mother and fetus(es) under aseptic conditions (5) and all observations were made on conscious animals at least four days after operation. Blood samples for analysis were obtained from the arterial catheters before feeding the ewes. Plasma insulin (6) and glucagon were measured by radioimmunoassay, using ovine insulin and porcine glucagon standards and two different antibodies for the glucagon assays; one (7) reacted with glucagon-like material of intestinal origin (enteroglucagon) as well as the pancreatic hormone, the other, Unger 30K,

was specific for pancreatic glucagon. The term "glucagon" is used
for the pancreatic hormone and GLI for the glucagon-like immunoreactivity measured with the non-specific antibody.

WELL-OXYGENATED FETUSES

The pregnant sheep has an exceptionally large materno-fetal gradient
for both oxygen and glucose (table 1 and figure 1A). Despite
maternal arterial PO_2 and plasma glucose levels comparable to those
in other species, the normal fetal arterial PO_2 in late gestation is
little more than 20 mm Hg and the fetal plasma glucose is only 15 -
25 mg/100 ml. Although the fetal plasma lactate is higher than the
maternal level, this is likely to be a consequence of deficient
gluconeogenesis (8) rather than hypoxia, for similar values are seen
on the day of birth in lambs with adult arterial PO_2 levels. Despite
the low plasma glucose, the fetal plasma insulin is in the same range
as the maternal level and the fetal plasma glucagon and GLI (figure 1B)
are usually lower. Since the sheep placenta is impermeable to
insulin and glucagon (9,10), this suggests that the sensitivity of

Table 1. Arterial PO_2, PCO_2 and pH
in five chronically-catheterized pregnant
ewes and their seven fetuses. Means $\pm$
S.E.M. Data supplementing figure 1.

	Gestation days	PO_2 mm Hg	PCO_2 mm Hg	pH
Mother:				
	118	106 $\pm$ 0	36.0 $\pm$ 2.0	7.43 $\pm$ 0.02
	123	95 $\pm$ 9.8	36.3 $\pm$ 1.5	7.48 $\pm$ 0.03
	128	104 $\pm$ 9.7	33.7 $\pm$ 1.2	7.49 $\pm$ 0.02
	133	94 $\pm$ 7.4	36.0 $\pm$ 1.4	7.46 $\pm$ 0.03
	137	90 $\pm$ 12.0	35.5 $\pm$ 0.5	7.53 $\pm$ 0.01
Fetus:				
	118	23.0 $\pm$ 2.0	51.0 $\pm$ 2.5	7.34 $\pm$ 0.04
	123	22.3 $\pm$ 2.0	50.8 $\pm$ 2.3	7.34 $\pm$ 0.01
	128	22.0 $\pm$ 1.9	49.5 $\pm$ 2.6	7.33 $\pm$ 0.02
	133	20.6 $\pm$ 0.4	50.0 $\pm$ 2.1	7.33 $\pm$ 0.01
	137	18.0 $\pm$ 0.9	45.3 $\pm$ 2.4	7.33 $\pm$ 0.02

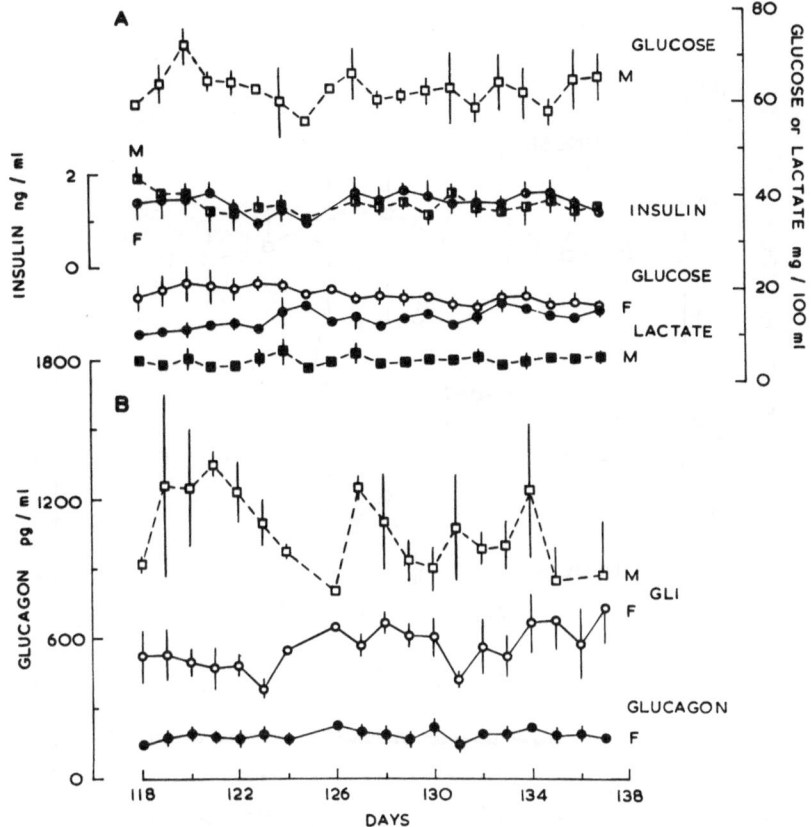

Figure 1: A. Maternal (M) and fetal (F)
plasma glucose (□ , ○), insulin (▤ ,◑) and
lactate (▨ , ●) in five chronically-catheter
-ized normoglycaemic ewes and their seven
fetuses.   All fetuses survived in good condition (table 1) for at least two weeks post-
operatively. B.   Maternal and fetal plasma
total glucagon-like immunoreactivity (GLI,
▢ , ○) and fetal plasma pancreatic glucagon
(●) in the same animals. Means ± S.E.M.
Term is at 1·47 days' gestation.

the fetal pancreas to glucose may differ from that of the adult.

This impression was confirmed when physiological amounts of

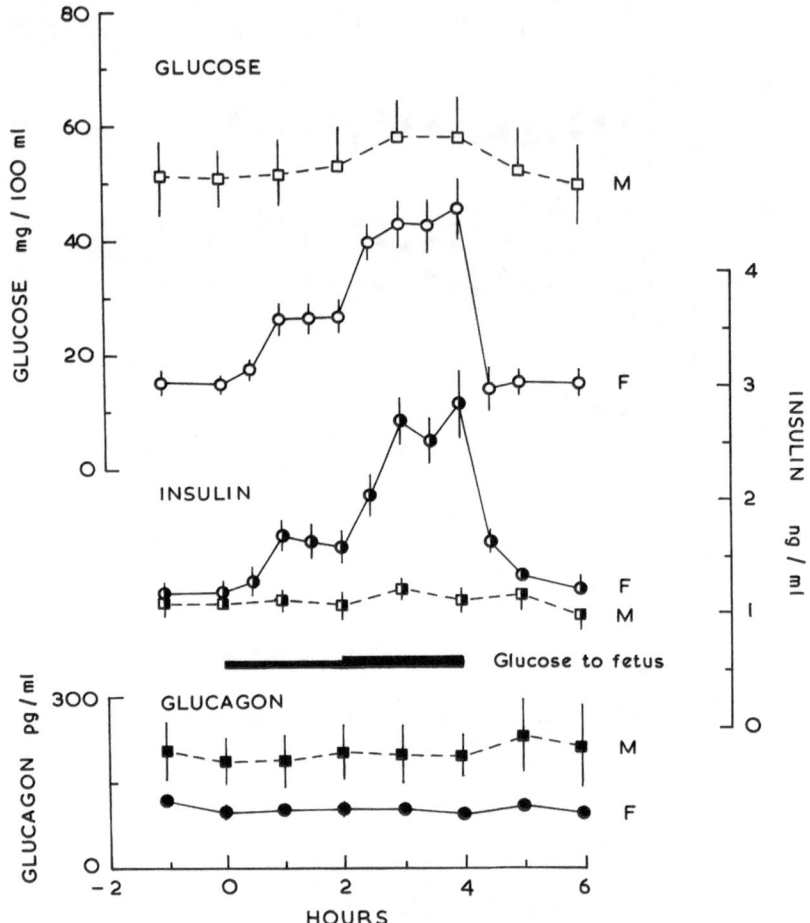

Figure 2: Maternal (M) and fetal (F)
plasma glucose (□,o), insulin (▯ , ◑)
and pancreatic glucagon (■ , ●) during
glucose infusion. Glucose was administered intravenously to the fetal lambs
for two hours at 27 mg/min followed by
two hours at 55 mg/min. Means ± S.E.M.
from four experiments at 124-143 days'
gestation.

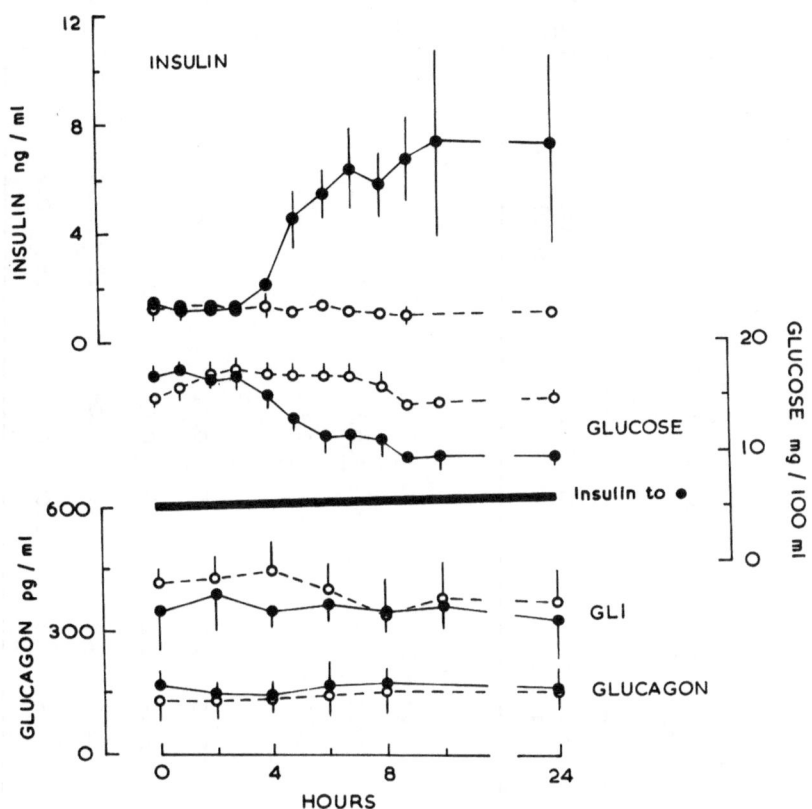

Figure 3: Fetal plasma insulin, glucose,
total glucagon-like immunoreactivity (GLI)
and pancreatic glucagon during insulin
infusion.   Insulin was administered intravenously, without first filling the dead
space of the catheter, to one of twins (●)
using the other lamb (O) as a control.
Means + S.E.M. from six sets of twins at
119-136 days' gestation.

glucose were administered to the fetus (figure 2).   Although the
fetal plasma glucose was kept below that in the mother, the fetal
plasma insulin rose to nearly three times the maternal level for a
three-fold rise of fetal plasma glucose.   The fetal plasma glucagon

Table 2. Plasma insulin, pancreatic glucagon and the insulin : glucagon ratio in normoglycaemic and hypoglycaemic pregnant ewes and fetal lambs. Means ± S.E.M.

	No. of samples (animals)	Glucose mg/100 ml	Insulin ng/ml	Glucagon pg/ml	Insulin / Glucagon
Mother:					
Normoglycaemic	20(4)	56.9 ± 1.8	1.38 ± 0.25	177 ± 9.5	7.8 ± 0.5
Hypoglycaemic	18(2)	30.4 ± 1.6	0.76 ± 0.03	148 ± 6.1	5.3 ± 0.3
	20(4)	37.8 ± 1.4	0.84 ± 0.04	349 ± 17.2	2.5 ± 0.2
Fetus:					
Normoglycaemic	28(6)	22.5 ± 0.5	1.39 ± 0.11	174 ± 8.8	8.3 ± 0.7
	39(12)	16.9 ± 0.2	1.44 ± 0.08	165 ± 9.1	9.8 ± 0.6
Hypoglycaemic	26(11)	12.2 ± 0.3	1.23 ± 0.09	135 ± 12.0	10.1 ± 0.8
	39(10)	7.5 ± 0.3	1.00 ± 0.09	118 ± 10.4	9.9 ± 0.6

remained at half the maternal level. When insulin was infused (figure 3), the plasma glucose fell below 10 mg/100 ml., demonstrating that fetal glucose uptake responded to physiological changes of plasma insulin, but again there was no change of plasma glucagon or GLI. In adult sheep acute insulin hypoglycaemia provokes a rise of plasma glucagon and GLI (11, 12).

Pregnant sheep also differ from non-ruminants in being highly susceptible to fasting hypoglycaemia. If the mother reduces her food intake, her plasma glucose falls below 40 mg/100 ml. and the fetal plasma glucose drops below 10 mg/100 ml. As shown in table 2, both mother and fetus lower their plasma insulin in response to fasting hypoglycaemia, but whereas most ewes raise their plasma glucagon, that of the fetus tends to fall. The net result is that while the plasma insulin : glucagon ratio falls substantially in hypoglycaemic ewes, there is no such change in the fetus; the ratio remains slightly higher than in the normoglycaemic adult.

In contrast, plasma GLI fell in fasting ewes and appeared to be more closely related to food intake than to plasma glucose (table 3). If the values for pancreatic glucagon in table 2 are subtracted from plasma GLI, the remainder, which probably represents plasma entero-glucagon, falls from about 970 pg equiv/ml in ewes on a high food

Table 3. Plasma total glucagon-like immunoreactivity (GLI) in normoglycaemic and hypoglycaemic pregnant ewes and fetal lambs. Means $\pm$ S.E.M.

	No. of samples (animals)	Food intake* g/day	Glucose mg/100 ml	GLI pg equiv/ml
Mother:				
Normoglycaemic	26(4)	1381 + 44	63.5 + 1.6	1145 + 57
	12(4)	443 + 80	51.3 + 9.4	765 + 30
Hypoglycaemic	5(1)	210 + 57	39.7 + 2.5	810 + 15
Fetus:				
Normoglycaemic	28(6)	—	22.5 + 0.5	577 + 30
	26(7)	—	17.2 + 0.3	598 + 47
Hypoglycaemic	9(6)	—	12.9 + 0.6	684 + 85
	9(2)	—	7.5 + 0.8	622 + 68

*The weight of concentrates consumed the previous day.

intake to less than half this value on a restricted food intake.
It is of some interest that, in a series of fetuses where glucagon
and GLI were determined on the same samples, the values for "plasma
enteroglucagon" were similar to those in fasting ewes.   The fetal
plasma GLI and "enteroglucagon" did not change with fetal plasma
glucose.

EFFECTS OF HYPOXIA

Fetal and newborn animals have an unusual ability to survive hypoxia
and the fetal lamb is no exception (13).   If the fetal arterial PO_2
is lowered below 18 mm Hg, by giving the ewe a low-oxygen mixture to
breathe, the fetus usually survives for several hours, large amounts
of lactic acid accumulate in the blood and tissues and there is a fall
of arterial pH.   The rate at which these changes occur depends not
only on the arterial PO_2, but also on the fetal plasma glucose concentration during the hypoxic episode (14).

Hypoxia readily suppresses insulin release from the fetal
pancreas and, if sufficiently severe, stimulates glucagon release.
In figure 4, the plasma glucose was raised by infusion from 10 to 70
mg/100 ml in one of twin fetal lambs.   Both twins were then made progressively more hypoxic by giving the ewe two oxygen mixtures to
breathe.   Whereas the plasma insulin fell immediately in both lambs,
while the ewe was breathing 8.5% O_2 and the fetal arterial pH was still
above 7.32, there was no change of plasma glucagon until the ewe was
given 7% O_2 and the fetal pH fell below 7.29.   There was then an
eleven-fold rise of plasma glucagon in the more lactacidotic, glucose-
treated twin, and a five-fold rise in the control.   In both lambs
the plasma glucose rose continuously throughout the two hours' hypoxia,
reaching 124 mg/100 ml in the infused twin and 49 mg/100 ml in the
control.

Analagous results were obtained in a series of fourteen hypoxia
experiments on three pairs of twins.   Fetal plasma insulin was
depressed in every experiment, but the plasma glucagon (table 4) only
rose if the plasma lactate rose at a rate of more than 1 mg/100 ml/min
and the pH fell below 7.29.   The size of the response was related to

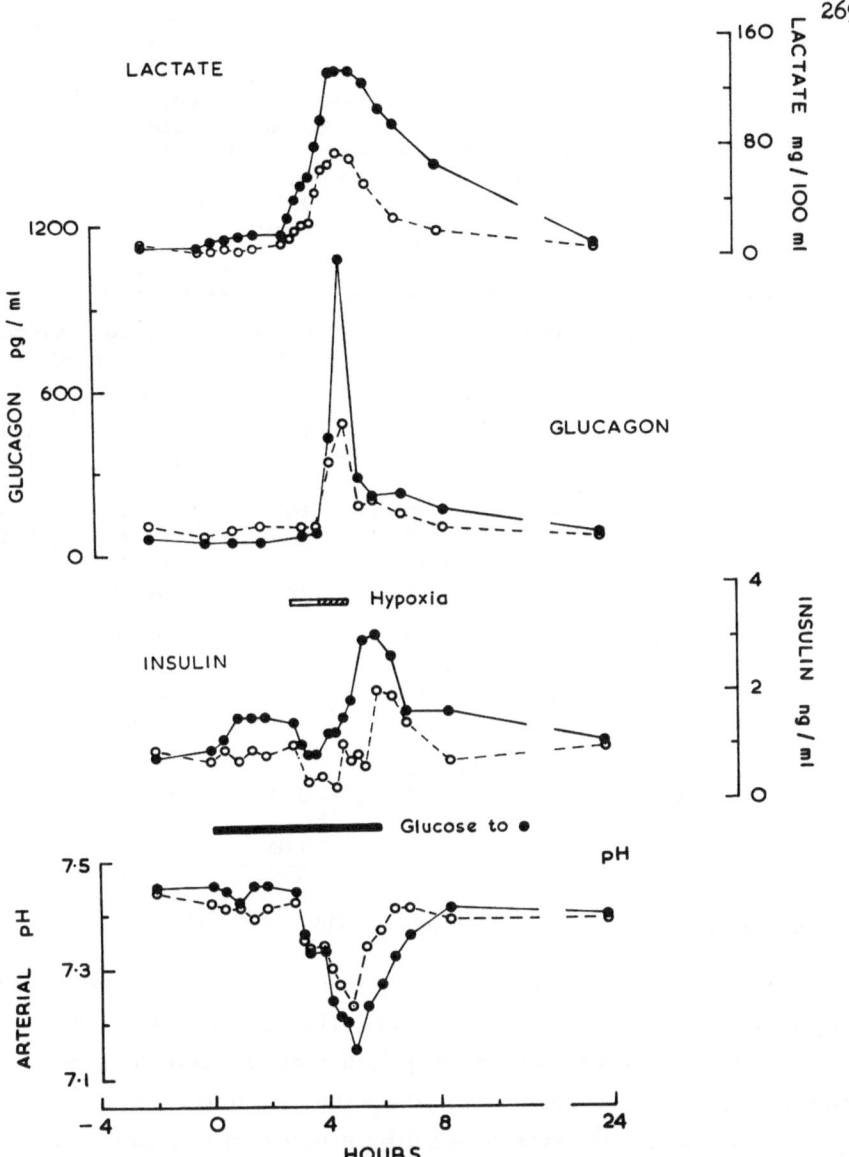

Figure 4: The effect of hypoxia on the plasma lactate, pancreatic glucagon, insulin and arterial pH of twin fetal lambs at 130 days' gestation.   One fetus (●) received glucose intravenously at 100 mg/min for six hours, the other (O) received saline acidified to pH5, that of the sterile 50% glucose solution.   After three hours both fetuses were made hypoxic by giving the ewe 8.5% O_2 + 3% CO_2 in nitrogen to breathe for one hour followed by 7% O_2 + 3% CO_2 in nitrogen for a second hour.

Table 4. The arterial PO_2 and pH,
plasma pancreatic glucagon, lactate
and glucose in three pairs of twin
fetal lambs after 60 min. hypoxia.
The ewe was breathing 8.5 or 7%
oxygen + 3% CO_2 in nitrogen.

Fetus	Gestation days	PO₂ mm Hg	pH	Glucagon pg/ml	Lactate	Glucose
					mg/100 ml	
174A	126	12	7.25	120	86.6	18.5
B	126	16	7.15	125	111.6	25.6
76A	128	13	7.39	100	34.7	14.5
	130	13	7.33	100	60.1	96.4*
	132	13	7.23	700	80.0	45.2
	135	14	7.28	420	86.0	31.0
76B	128	12	7.37	120	31.3	41.0*
	130	14	7.34	125	26.5	25.7
	132	15	7.19	270	102.0	86.2*
	135	12	7.29	120	58.8	24.4
153A	132	17	7.32	100	63.4	18.5
	134	17	7.05	320	124.0	95.0*
153B	132	16	7.22	110	99.4	57.4*
	134	16	7.19	380	98.6	25.0

*The fetus received glucose intravenously throughout the experiment.

the extent of the lactacidaemia in each individual lamb. The absence
of any response in the two youngest lambs (126 days' gestation),
despite a profound lactacidaemia, suggests that the threshold for
glucagon release may decrease with advancing gestational age.

This impression was strengthened by studies on spontaneously
hypoxic fetal lambs. Some fetuses fail to recover completely from
the effects of surgery. They are characterized by low arterial PO_2
and plasma glucose levels, slightly raised but stable lactate levels,
and a normal or raised arterial pH. If undisturbed, they survive
for days or even weeks, but glucose administration provokes lactacidaemia and fetal death. As shown in figure 5, where the plasma glucose
was raised to 75 mg/100 ml in one of hypoxic twins, glucose fails
to elicit an insulin response in hypoxic fetuses. The basal plasma

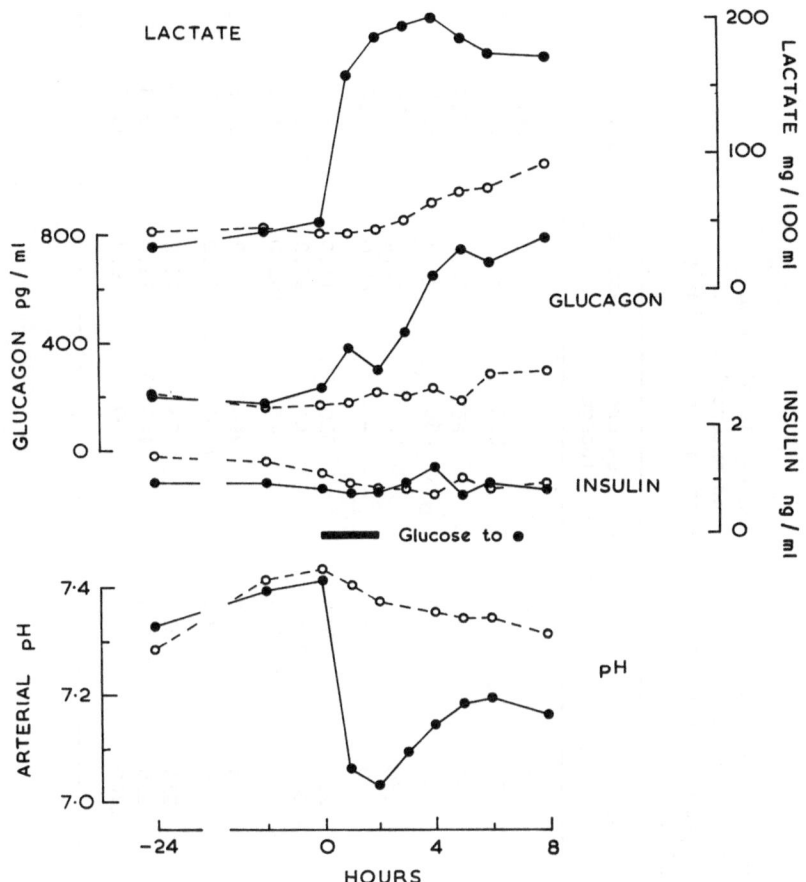

Figure 5: Plasma lactate, pancreatic glucagon,
insulin and arterial pH in hypoxic twin fetal
lambs at 123-124 days' gestation.   One twin (●)
received glucose intravenously at 55 mg/min for
two hours, the control (○) received acid saline.
The resting arterial PO_2 in both twins was 11 - 13
mm Hg.

glucagon was slightly raised (x 2), and the lactacidaemia was
accompanied by a slow sustained rise of plasma glucagon to four
times the resting level, despite the young age of the fetus (124
days' gestation).   A rise of plasma glucagon was also seen in young,

Table 5. The arterial PO$_2$ and pH, plasma pancreatic glucagon, total glucagon-like immunoreactivity (GLI) and enteroglucagon (by difference), plasma lactate and glucose in six spontaneously hypoxic fetal lambs.

Fetus	Gestation days	PO$_2$ mm Hg	pH	Glucagon	GLI pg equiv/ml	Entero-glucagon	Lactate mg/100 ml	Glucose
111A	116	21	7.35	200	300	100	22.0	26.6
	117	-	-	470	1280	810	87.2	11.2
111B	118	17	7.36	95	380	285	16.6	18.4
	119	11	7.18	250	700	450	76.7	11.5
127A	119	14	7.37	150	-	-	22.4	5.3
	120	11	7.25	430	-	-	186.0	7.2
102B	123	19	7.29	290	800	510	11.0	6.5
	124	17	7.27	300	1200	900	27.2	5.7
	125	21	7.02	> 2000	> 2000	-	151.0	8.4
165A	124	-	7.30	115	320	205	28.9	19.3
	125	-	6.98	900	2100	1200	164.0	8.7
36A	134	18	7.31	215	350	135	16.8	24.0
	135	16	7.30	355	460	105	29.0	15.3
	136	13	7.24	1200	2100	900	76.7	11.6

spontaneously-deteriorating fetuses (table 5); in relation to the
lactacidaemia, the response was largest in the oldest fetus (135 –
136 days' gestation). The rise of plasma glucagon was usually
accompanied by, and sometimes preceded by, a rise of plasma GLI larger
than could be accounted for by the change of plasma glucagon. The
calculated rise of plasma entero-glucagon appeared to be independent
of gestational age.

A relationship between plasma glucagon and lactacidaemia was
also seen post-natally (figure 6). Two lambs catheterized two weeks
before delivery had respiratory problems on the day of birth (◑, ●);
the plasma glucagon rose and fell with the plasma lactate. A third
lamb (X) was treated with oxygen by face mask. It developed a
respiratory acidosis, but there was no rise of glucagon or lactate;
the levels remained close to the pre-natal values and those in a
clinically-normal lamb (O) catheterized four weeks earlier. Plasma
glucose was variable and unrelated to plasma glucagon, plasma insulin
was always low.

DISCUSSION

These observations confirm and extend earlier reports that glucose
administration and hypoglycaemia had no effect on plasma GLI and
glucagon in fetal lambs, rats and rhesus monkeys (12, 15 – 19). The
changes of fetal plasma insulin in response to hypo- and hyperglycaemia
have also been described before (18, 20 – 22), the effect of insulin
on net fetal glucose uptake has been demonstrated in fetal lambs
(23, 24), and the ability of the fetal lamb pancreas to release insulin
at much lower glucose concentrations than after birth has been shown
in vitro (25).

The effects of hypoxia have not been explored before. The readi-
ness with which this inhibits insulin release may explain why many
workers failed to demonstrate the effect of glucose on fetal plasma
insulin (for references see 26). The mechanism is likely to involve
catecholamines which are released into the fetal circulation during
hypoxia and are known to suppress insulin secretion (16, 18, 27, 28).
Catecholamines may also be responsible for the rise of plasma glucagon,
for their infusion stimulates glucagon release in fetal lambs and rats

274

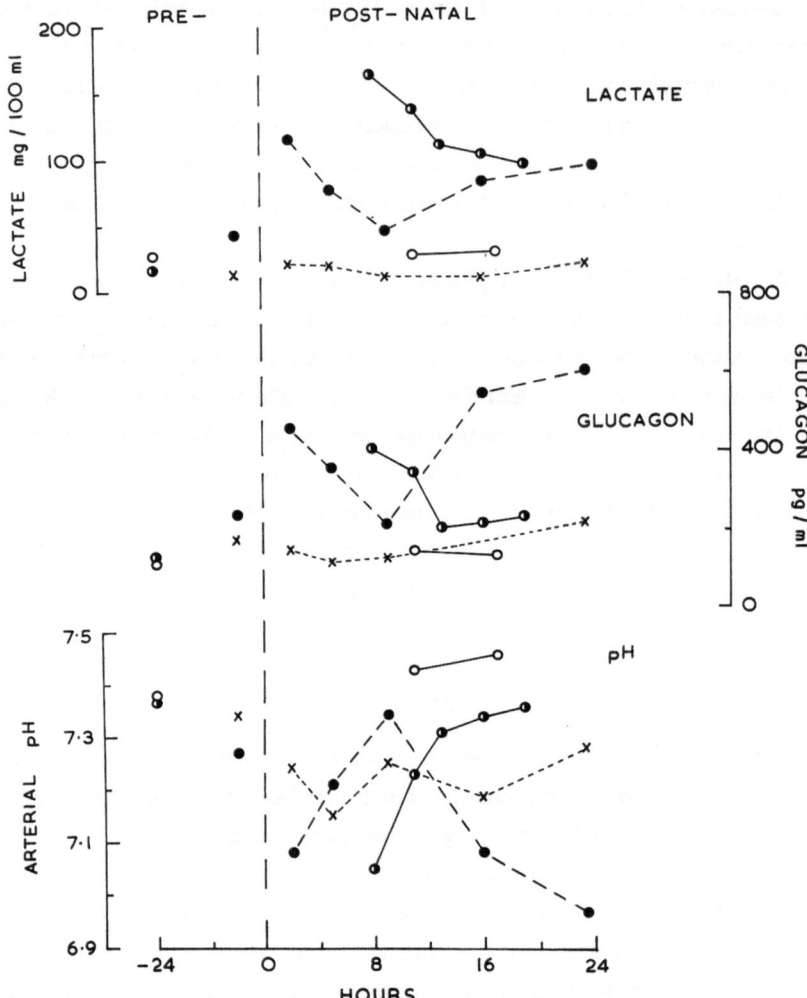

Figure 6: Plasma lactate, pancreatic
glucagon and arterial pH before and after
birth in four lambs delivered spontaneously
at 145 (O) and 140 (◑) days' gestation
or by Caesarean section at 134 days (● , X).
For details see text.

(16, 18). But the hypoxic threshold for glucagon release is clearly higher than for insulin suppression, and autonomic stimulation via the pancreatic nerves may be necessary for glucagon release in vivo (29 - 31). The putative decrease in this threshold towards term could be due to increased central sensitivity to the effects of hypoxia or to changes in the innervation of the pancreas.

Plasma glucagon is also high in newborn rats (32) and rises on cutting the umbilical cord of fetal lambs (33). As in the present study, high glucagon levels in the newborn rat were associated with high lactate levels and fell in parallel with lactate, suggesting that hypoxia was the stimulus for its release. Plasma glucagon in cord blood from hypoxic newborn babies is higher than in that from normal babies (34) and, judging from table 4, the transitory changes of plasma lactate and pH following normal delivery of fetal lambs (35) indicate a degree of hypoxia sufficient to trigger a rise of plasma glucagon in even the least affected lamb; the pH always fell below 7.25 and the plasma lactate rose by at least 25 mg/100 ml in the first five minutes.

The metabolic consequences of these endocrine changes are likely to benefit the fetus. Fetuses of all species accumulate large amounts of glycogen for use after birth (36). Insulin promotes glucose incorporation into fetal hepatic, cardiac and skeletal muscle glycogen (37 - 39), whereas glucagon stimulates fetal hepatic glycogen-olysis in late gestation and causes a rise of fetal plasma glucose (8, 12, 18, 40 - 42). In fasting hypoglycaemia, the fall of fetal plasma insulin may be effective in decreasing peripheral glucose uptake, but the constancy of the insulin : glucagon ratio (table 2) will tend to preserve the glycogen stores. In severe hypoxia, however, the fetus survives at the expense of its glycogen reserves (13) and the depression of plasma insulin and rise of plasma glucagon will augment the effects of catecholamines in promoting glycogen breakdown.

In adults, glucagon also increases hepatic glucose production by stimulating gluconeogenesis. Although glucagon administration to fetal rats near term increases the activities of hepatic glucose-6-phosphatase and phosphoenolpyruvate carboxykinase (41), it is unlikely to affect gluconeogenesis until after birth (8), when this process

becomes of vital importance for the maintenance of plasma glucose.
For some years there has been speculation that the neonatal "glucagon
surge" could be responsible for the very large post-natal rise of
gluconeogenic enzymes which can be checked by injecting insulin or
glucose (41). Glucagon may also trigger the post-natal mobilization
of liver glycogen, for this is unaffected by adrenalectomy or
catecholamine-blocking agents.

The role of enteroglucagon is still unclear, even in adults, but
large amounts are present in the large and small intestines of fetal
lambs and human babies from early in gestation (16, 43). Unlike
pancreatic glucagon, it consists of at least two molecular species,
one of which may have functions similar to those of pancreatic glucagon (44). In newborn and adult, the plasma level rises after feeding
(44, 45, table 3) and may facilitate the post-prandial rise of plasma
insulin. Its ready release in hypoxic fetal lambs is intriguing and
deserves further investigation.

Further work is also needed to establish whether the well-known
opposing actions of insulin and glucagon on lipid and aminoacid
metabolism (4) can also operate in the fetus. The evidence available
suggests that insulin is an important determinant of fetal growth and
adiposity (46, 47), but these aspects of glucagon action have been
totally neglected.

This work was supported by the Medical Research Council. I am
greatly indebted to Dr. R. Assan and Mde. J. Boillot for performing
the glucagon assays and to many colleagues for surgical assistance.
Mrs. M. Ramage and Mrs. J. Inman gave valuable technical support.

REFERENCES

1. Silver, M, Fetal energy metabolism, In: Fetal Physiology and
    Medicine, Beard, RW and PW Nathanielsz (ed.), London,
    W.B. Saunders, 173-193, 1976
2. Shelley, HJ and GA Neligan, Neonatal hypoglycaemia. Br med Bull
    22 : 34-39, 1966
3. Shelley, HJ, Transfer of carbohydrate, In: Placental Transfer,
    Chamberlain, GVP and AW Wilkinson (ed.), Tunbridge Wells,
    Pitman Medical, 118-141, 1979
4. Exton, JH and CR Park, Interaction of insulin and glucagon
    in the control of liver metabolism, In: Handbook of Physiol-
    ogy, Section 7 : Endocrinology, Vol.1, Endocrine Pancreas,
    Steiner, DF and N Freinkel (ed.), Washington, American
    Physiological Society, 437-455, 1972

5.   Dawes, GS, HE Fox, BM Leduc, GC Liggins, and RT Richards,
     Respiratory movements and rapid eye movement sleep in the
     foetal lamb.   J Physiol 220 : 119-143, 1972
6.   Hales, CN and PJ Randle, Immunoassay of insulin with insulin-anti-
     body precipitate. Biochem J 88 : 137-146, 1963
7.   Assan, R, G Tchobroutsky, and M Derot, Glucagon radioimmunoassay
     : technical problems and recent data.   Horm Metab Res 3 :
     Suppl 1, 82-90, 1971
8.   Warnes, DW, RF Seamark, and FJ Ballard, The appearance of
     gluconeogenesis at birth in sheep.   Activation of the pathway
     associated with blood oxygenation. Biochem J 162 : 627-634,
     1977
9.   Alexander, DP, HG Britton, NM Cohen, and DA Nixon, The permeabili-
     ty of the sheep placenta to insulin.   Studies with the
     perfused placental preparation. Biol Neonate 21 : 361-368,
     1972
10. Alexander, DP, R Assan, HG Britton, and DA Nixon, Impermeability
     of the sheep placenta to glucagon.   Biol Neonate 23 : 391-402,
     1973
11. Arcus, AC, MJ Ellis, RD Kirk, DW Beavan, RA Donald, DS Hart, GW
     Holland, and C Redekopp, Studies with the auto transplanted
     ovine pancreas : glucagon and insulin secretion.   Aust J biol
     Sci 29 : 223-236, 1976
12. Alexander, DP, R Assan, HG Britton, E Fenton, and D Redstone,
     Glucagon release in the sheep fetus.
     1) Effect of hypo- and hyperglycaemia and arginine.   Biol
     Neonate 30 : 1-10, 1976
13. Dawes, GS, JC Mott, and HJ Shelley, The importance of cardiac
     glycogen for the maintenance of life in foetal lambs and new-
     born animals during anoxia.   J Physiol 146 : 516-538, 1959
14. Shelley, HJ, JM Bassett, and RDG Milner, Control of carbohydrate
     metabolism in the fetus and newborn. Br med Bull 31 : 37-43,
     1975
15. Fiser, RH, A Erenberg, MA Sperling, W Oh, and DA Fisher, Insulin-
     glucagon substrate interrelations in the fetal sheep.   Pediat
     Res 8 : 951-955, 1974
16. Bassett, JM, Glucagon, insulin and glucose homeostasis in the
     fetal lamb.   Ann Rech Vét 8 : 362-373, 1977
17. Schreiner, RL, PA Nolen, PW Bonderman, HC Moorehead, EL Gresham,
     JA Lemons, and MB Escobeda, Fetal and maternal hormonal resp-
     onse to starvation in the ewe. Pediat Res 14 : 103-108, 1980
18. Girard, JR, A Kervran, E Soufflet, and R Assan, Factors affecting
     the secretion of insulin and glucagon by the rat fetus.
     Diabetes 23 : 310-317, 1974
19. Chez, RA, DH Mintz, MF Epstein, AR Fleischmann, GK Oakes, and DL
     Hutchinson, Glucagon metabolism in non-human primate pregnancy.
     Amer J Obstet Gynec 120 : 690-694, 1974
20. Bassett, JM and D Madill, The influence of maternal nutrition on
     plasma hormone and metabolite concentrations of foetal lambs.
     J. Endocr 61 : 465-477, 1974
21. Bassett, JM and D Madill, Influence of prolonged glucose infusions
     on plasma insulin and growth hormone concentrations of foetal
     lambs.   J Endocr 62 : 299-309, 1974

22. Philipps, AF, BS Carson, G Meschia, and FC Battaglia, Insulin secretion in fetal and newborn sheep. Amer J Physiol 235 : E467–E474, 1978

23. Simmons, MA, MD Jones Jr, FC Battaglia, and G Meschia, Insulin effect on fetal glucose utilization. Pediat Res 12 : 90–92, 1978

24. Carson, BS, AF Philipps, MA Simmons, FC Battaglia, and G Meschia, Effects of a sustained insulin infusion upon glucose uptake and oxygenation of the ovine fetus. Pediat Res 14 : 147–152, 1980

25. Bassett, JM, D Madill, DH Nicol, and GD Thorburn, Further studies on the regulation of insulin release in foetal and post-natal lambs : the role of glucose as a physiological regulator of insulin release in utero, In: Foetal and Neonatal Physiology, Comline, RS, KW Cross, GS Dawes, and PW Nathanielsz (ed.), London, Cambridge University Press, 351–359, 1973

26. Shelley, HJ, The use of chronically catheterized foetal lambs for the study of foetal metabolism, In: Foetal and Neonatal Physiology, Comline, RS, KW Cross, GS Dawes, and PW Nathanielsz (ed.), London, Cambridge University Press, 360–381, 1973

27. Jones, CT and RO Robinson, Plasma catecholamines in foetal and adult sheep. J Physiol 248 : 15–33, 1975

28. Jones, CT and JWK Ritchie, The metabolic and endocrine effects of circulating catecholamines in fetal sheep. J Physiol 285 : 395–408, 1978

29. Marliss, EB, L Girardier, J Seydoux, CB Wollheim, Y Kanazawa, L Orci, AE Renold, and D Porte Jr, Glucagon release induced by pancreatic nerve stimulation. J clin Invest 52 : 1246–1259, 1973

30. Bloom, SR, AV Edwards, and NJA Vaughan, The role of the sympathetic innervation in the control of plasma glucagon concentration in the calf. J Physiol 233 : 457–466, 1973

31. Bloom, SR, AV Edwards, and NJA Vaughan, The role of the autonomic innervation in the control of glucagon release during hypoglycaemia in the calf. J Physiol 236 : 611–623, 1974

32. Girard, JR, GS Cuendet, EB Marliss, A Kervran, M Rieutort, and R Assan, Fuels, hormones and liver metabolism at term and during the early postnatal period in the rat. J clin Invest 52 : 3190–3200, 1973

33. Grajwer, LA, MA Sperling, J Sack, and DA Fisher, Possible mechanisms and significance of the neonatal surge in glucagon secretion : studies in newborn lambs. Pediat Res 11 : 833–836, 1977

34. Johnston, DI and SR Bloom, Plasma glucagon levels in the term human infant and effect of hypoxia. Archs Dis Childh 48 : 451–454, 1973

35. Comline, RS, and M Silver, The composition of foetal and maternal blood during parturition in the ewe. J Physiol 222 : 233–256, 1972

36. Shelley, HJ, Glycogen reserves and their changes at birth and in anoxia. Br med Bull 17 : 137–143, 1961

37. Eisen, HJ, ID Goldfine, and WH Glinsman, Regulation of hepatic glycogen synthesis during fetal development : roles of hydrocortisone insulin and insulin receptors. Proc Nat Acad Sci 70 : 3454–3457, 1973

38. Clark, CM Jr, Characterization of glucose metabolism in the isolated rat heart during fetal and early neonatal development. Diabetes 22 : 41-49, 1973

39. Bocek, RM, MK Young, and CH Beatty, Effect of insulin and epinephrine on the carbohydrate metabolism and adenylate cyclase activity of rhesus fetal muscle. Pediat Res 7 : 787-793, 1973

40. Bassett, JM and GD Thorburn, The relguation of insulin secretion by the ovine fetus in utero. J Endocr 50 : 59-74, 1971

41. Girard, JR, D Caquet, D Bal, and I Guillet, Control of rat liver phosphorylase, glucose-6-phosphatase and phosphoenolpyruvate carboxykinase activities by insulin and glucagon during the perinatal period. Enzyme 15 : 272-285, 1973

42. Chez, RA, DH Mintz, EO Horger III, and DL Hutchinson. Factors affecting the response to insulin in the normal subhuman pregnant primate. J clin Invest 49 : 1517-1527, 1970

43. Assan, R and J Boillot, Pancreatic glucagon and glucagon-like material in tissues and plasma from human fetuses 6-26 weeks old. Pathologie Biologie 21 : 149-157, 1973

44. Unger, RH, Circulating pancreatic glucagon and extrapancreatic glucagon-like materials, In: Handbook of Physiology, Section 7 : Endocrinology, Vol 1, Endocrine Pancreas, Steiner, DF and N Freinkel (ed.), Washington, American Physiological Society, 529-544, 1972

45. Sperling, MA, PV Delamater, D Phelps, RH Fiser, W Oh, and DA Fisher, Spontaneous and amino-acid-stimulated glucagon secretion in the immediate post-natal period : Relation to glucose and insulin. J clin Invest 53 : 1159-1166, 1974

46. Hill, DE, Insulin and fetal growth. Prog clin biol Res 10 : 127-139

47. Liggins, GC, The drive to fetal growth, In: Fetal Physiology and Medicine, Beard, RW and PW Nathanielsz (ed.), London, W.B. Saunders, 254-270, 1976

EFFECTS OF INSULIN AND GLUCAGON ON THE MOTHER AND IN THE FETUS

R.D.G. Milner

INTRODUCTION

The title of this review looks tidy. It suggests equal
attention will be given to the hormones insulin and
glucagon and this will be permutated against the mother
and the fetus. But when the title is taken in the context
of the overall theme of the book another picture emerges.
The placenta is impermeable to either hormone (1,2) and
when given to the mother effects on the fetus are indirect
and result from changes in metabolite delivery. In contrast insulin and glucagon have direct effects on the fetus
and it is the fetus that is the focal point of the
discussion.

This review will therefore concentrate on the fetus at
the expense of the mother. Emphasis will be placed not
only on the administration of exogenous hormones to mother
or fetus but to conditions of over- and undersecretion of
either hormone by the fetus.

THE MOTHER

The majority of pregnant diabetic women are treated with
insulin, but it is rare for a pregnant woman or animal to
be given glucagon. Since glucagon administration has
little if any therapeutic relevance and endogenous maternal
hyperglucagonemia in pregnancy is not recognised as a
clinical entity, this facet of the review will not be
pursued.

Carbohydrate metabolism in pregnancy has been reviewed in detail (3,4). It is accepted that pregnancy
per se is diabetogenic and that maternal insulin resistance
or insulinopenia during pregnancy results in maternal and
fetal metabolic derangement with increased fetal morbidity

and mortality. Questions which are topical revolve around how rigorously maternal metabolism, as reflected in blood glucose profiles, should be normalised. It has been claimed that insulin therapy to maximally tolerated doses will correct perinatal mortality and morbidity (5). How vigorous should we be in screening for gestational diabetes? Should women with minor glucose intolerance during pregnancy be treated dietetically or with insulin? What are the implications for man of the observation that the female offspring of rats with reduced carbohydrate tolerance manifest gestational diabetes which becomes severer in succeeding generations (6,7)?

These questions illustrate clinical priorities; the biological aspects of the review are broader and best considered by a detailed review of the fetus.

THE FETUS

The administration of insulin or glucagon to the fetus or the effects of changes in circulating levels of either hormone must be analysed with reference to the stage of fetal development that the event occurs. The responsiveness of the fetal islet to stimuli of insulin or glucagon secretion has been reviewed recently (8) and is not considered here.

In what follows it is assumed that insulin and glucagon in the fetus are the same molecular species as after birth and that neither hormone is bound by any "carrier" molecule. It will be seen that effects of each hormone depend importantly on the ontogeny of hormone receptors which differ between organs and between species.

Effects of Insulin and Glucagon in the Whole Fetus

A variety of effects have been described for injecting insulin or anti-insulin serum into rat fetuses. Early reports stated that exogenous insulin injection into the 20 day rat fetus had no effect on the incorporation of ^{14}C from glucose into lipid, glycogen or protein or ^{3}H into fatty acids. Contrary results were obtained subsequently which showed that insulin injection into rat fetuses at term increased the incorporation of ^{14}C-glucose into glycogen and that anti-insulin serum produced the opposite effect (9,10,11). It is possible that confusion could have arisen because of the high endogenous plasma insulin concentration at the time of exogenous insulin injection. If endogenous insulin was causing a maximum biological action it might not be possible to show a further effect by the injection of exogenous insulin. The action of anti-insulin serum suggests that this might have been the case.

In the last four days of intrauterine life of the rat both plasma insulin and glucagon concentrations were found to rise several fold, but with the preservation of an insulin/glucagon molar ratio of more than 10. Immediately after birth there was a precipitous fall in plasma insulin and a steep rise in plasma glucagon with a consequent drop in the insulin/glucagon ratio to 1 which was maintained during fasting for 16h (12). If a maternal fast of 96h was imposed immediately before term, the effect on the fetus was to lower blood glucose and gluconeogenic substrate and to raise blood ketone bodies (13). Plasma insulin fell and glucagon rose, anticipating the changes that occur normally in the immediate postnatal period. There was an increase in activity of key gluconeogenic enzymes in fetal liver of starved pregnancies (vide infra) suggesting that maternal starvation might induce precocious fetal gluconeogenesis at the price of impaired fetal growth.

What experiments of nature may be taken as clinical examples of the effects of insulin and glucagon on fetal growth? Much of the interest in the topic stems from the hyperinsulinemic overweight infant of a diabetic mother (IDM). In the extensive literature on that subject there has been little evidence of scientific rigour in analysing what types of cellular development comprise body weight. The characteristic IDM is overweight mainly due to excess lipid and there is a modest and arguable increase in cell size and number of the lean body mass. Subcutaneous fat in

normal infants and IDM develops in the last 10-12 weeks of
intrauterine life, yet it is inferred that the pancreatic
islet hyperplasia, β cell hypertrophy and hyperinsulinemia
antecedes this period of development. If structural and
functional β cell pathology does exist in the IDM during
the second trimester we may postulate that abnormal adipose
development occurs because (?pre-) adipocytes develop
insulin receptors at 28-30 weeks gestation. It must not be
forgotten however that there is little direct information
on pancreatic morphology of IDM or circulating plasma
insulin levels before 28 weeks gestation. If, as has been
suggested, glucose becomes a secretogogue for the human β
cell at about 28 weeks gestation, and fetal hyperglycemia
is the key to the pathology of IDM, then hyperinsulinemia
during the last 12 weeks of intrauterine life alone might
account for pathological fat accumulation.

The converse has been argued for the infant with
transient neonatal diabetes who is hypoinsulinemic and
certainly lacks subcutaneous fat as judged clinically.
But no clinical or scientific measurement has been made of
the lean body mass of these babies. Two siblings with
pancreatic agenesis have been reported (14) and in one
there was clear, if anecdotal, evidence of a reduction in
hepatic cellularity and muscle protein/DNA ratio.

Attempts to mimic diabetic pregnancy or to
"insulinectomise" the fetus have had very mixed results
largely because of the complexity of the experimental model.
It has been difficult to produce a consistently overweight
fetus associated with drug-induced maternal diabetes and
the techniques employed to deprive the fetus of β cell
function have been too imprecise to allow an answer to the
question of what action insulin plays in determining fetal
growth.

If a chronically catheterised sheep fetus is infused
with insulin to produce a rise in plasma concentration
within the physiological range there is a fall in blood
glucose and increase in glucose utilisation (15). Many of
the conflicting results in the early sheep experiments

arose from a failure to appreciate the need to work with a
chronic unstressed preparation. The failure to demonstrate
a fall in fetal blood glucose following insulin infusion
into the rhesus monkey fetus at 0.9 gestation (16) is open
to the same criticism. Alternatively the kinetics of
glucose transport across the monkey placenta may be such
that it was difficult to influence fetal blood glucose in
this way. Plasma free fatty acid levels fell consistently
following injection of insulin into the fetus.

Glucagon infused into newborn infants on the third
day of life produced a hypoaminoacidemia similar to that
seen in the adult (17). The gluconeogenic amino acids
alanine, glycine, proline and glutamine plus arginine
accounted for 60% of the change. On the first day of life
an attenuated response to glucagon was seen in normal and
small-for-dates infants but IDM were totally insensitive.
These results are in keeping with the idea that glucagon
is gluconeogenic in the neonate and that glucagon and
insulin have opposite actions on gluconeogenesis.

A clue to the role of glucagon in the human perinatal
period comes from a detailed case study of an infant with
isolated glucagon deficiency (18). The patient had a
birth weight of 3.81 kg at 42 weeks gestation and presented
with persistent hypoglycemia. Investigations revealed
normal insulin secretion, no plasma glucagon response to
hypoglycemia or alanine infusion and severe impairment of
gluconeogenesis. The condition appeared to be familial
and an autosomal recessive inheritance was suggested.
There was nothing to suggest a role for glucagon in
promoting fetal growth.

The futility of seeking a simple metabolic-hormonal
interaction for glucagon is best appreciated if the human
studies are reviewed in the light of animal experiments.
Studies on sheep have suggested that the immediate neonatal surge in plasma glucagon may be causally related to
severance of the umbilical cord (19). Lambs were delivered
into room air and the umbilical cord left intact and
covered with warm saline so that the blood supply to the

fetus was undisturbed. The fetal face was covered with a warm saline filled glove. No change in plasma glucagon occurred until the cord was cut 60 min later when a five fold rise occurred in 15 min. This was followed 30 min later by a large rise in plasma FFA. The authors suggested that both events were due to a surge in catecholamine release triggered by cutting the umbilical cord.

In a later study lambs aged 1-3 days were infused with somatostatin which suppressed endogenous insulin and glucagon release (20) and led to a fall in plasma glucose. From the results of infusion with exogenous insulin or glucagon the authors concluded that glucagon is a major hormone for maintaining blood glucose during short term neonatal fasting. But there is evidence showing that glucagon is not the only, and possibly not the primary, activator of neonatal gluconeogenesis. Gluconeogenesis in the newborn lamb has been demonstrated within 2 and 4 min of birth at term and to be delayed in two prematurely delivered lambs (139 and 144 days) until breathing was established and the blood fully oxygenated (21). Glucagon infusion at 131 or 144 days into two fetuses caused a rise in blood glucose but no gluconeogenesis. The authors proposed that oxygen availability initiates gluconeogenesis in the newborn lamb. Oxygen or glucagon, or both? The question will not be resolved until simultaneous measurements of the relevant variables are made in one set of experiments.

Effects of Insulin and Glucagon on Fetal Liver

Rat fetuses injected with insulin at term became hypoglycemic and formed more hepatic glycogen (9). Conversely fetuses injected with insulin antibody or glucagon became hyperglycemic and had decreased hepatic glycogen (9,10,22). These observations should not tempt the unwary to extrapolate. Although hepatocyte membrane glucagon receptors were demonstrated in 20 day rat fetuses the concentration was less than in neonatal or adult tissues (23). The result was due, not to increased glucagon degradation by

fetal tissue or a lack of adenyl cyclase in hepatocyte
membranes, but to a paucity of glucagon receptor sites.
Binding of ^{125}I glucagon by 15 day fetal hepatic membranes
was 1% of the adult level and 23% of that on day 21 (23).
In contrast insulin binding on day 15 was 11% of the adult
level and 45% of that on day 21. The fetal rat hepatocyte
is potentially more sensitive to the action of insulin
than glucagon. Both hormones become equipotent, that is
achieve their adult membrane binding characteristics by
30 days postnatally.

A comparison of the biochemical ontogeny of the rat
and human hepatocyte reveals several similarities
especially when enzymic activity is expressed per unit wet
liver weight and as a fraction of that found in the adult
(24). Of particular interest is the observation that some
enzymes which normally appear postnatally in the rat can
be evoked by the administration of glucagon to the fetus.
For example exogenous glucagon stimulates the activity of
tyrosine amino transferase, glucose-6-phosphatase,
phosphoenolpyruvate carboxykinase (PEPCK) and phosphorylase. The arginine synthetase system, the rate limiting
step in urea biosynthesis increases in the rat from a low
level at birth to peak activity on the second postnatal
day. Glucagon had no accelerating effect on this system
when injected into 19.5-21.5 day fetuses (25), but when
added to organ cultures of 18.5 day fetal liver caused a
doubling of arginine synthetase activity after 24h.
Insulin had no effect on the liver organ culture system.
It is probable that glucagon plays a part in the normal
enzymic ontogeny of glycogenolysis and gluconeogenesis
and is possible that precocious induction of these pathways may result if hyperglucagonemia is the consequence of
metabolic or drug induced alteration of the fetal environment. Greengard (24) suggests that in both rat and man,
thyroid hormone, glucocorticoids and glucagon may play a
part in controlling patterns of hepatic enzyme development.
In this context hormonal interaction must not be forgotten. Girard and his co-workers (26) confirmed the

induction of glucose-6-phosphatase and PEPCK by exogenous glucagon in 18.5-21.5 day-old rat fetuses. Phosphorylase and glycogen breakdown were not increased by glucagon before day 20.5, but if the fetuses were treated with hydrocortisone on day 18.5, phosphorylase activation by glucagon could be induced prematurely on day 19.5. Insulin administered at birth prevented the normal development of glucose-6-phosphatase or PEPCK but was without effect on hepatic phosphorylase activity and glycogen breakdown.

Villee was the first to demonstrate gluconeogenesis by human fetal liver in vitro but ethical and technical problems inhibited the further development of this technique until Schwartz and his colleagues performed a series of experiments in which human fetal liver explants were grown in organ culture. The tissue was obtained from fetuses of 5-25 weeks gestational age delivered by hysterotomy for therapeutic abortion. The incorporation of alanine into glucose and glycogen was stimulated by glucagon plus theophylline (27). Unfortunately the two agents were not tested independently and it is not possible to deduce if the effect was mediated by glucagon or theophylline. Insulin abolished the stimulation of gluconeogenesis by glucagon plus theophylline.

Insulin increased tissue glycogen accumulation from fetuses of 7 or more weeks gestation and glucagon depleted hepatic glycogen stores from 6 week fetuses (28). When both hormones were present in pharmacological and presumably maximum concentrations, the glucagon effect overrode the insulin effect. Further work led to the conclusion that the two hormones exerted their effect via the D-form of glycogen synthetase and did not influence phosphorylase activity.

When the human fetal liver organ culture was used to study amino acid uptake (29), α-amino isobutyrate accumulation was stimulated by glucagon, dibutyryl cyclic AMP or insulin. The effects of glucagon or dibutyryl cyclic AMP were not additive but the effect of insulin was additive to either, suggesting that insulin worked by a

different mechanism to glucagon and that the glucagon effect was transmitted via cyclic AMP. Glucagon sensitivity
was demonstrable at 6 weeks gestation (30 mm crown-rump
fetus) but insulin responsiveness was not demonstrated
below 48 mm crown-rump length. The physiological
significance of the same metabolic effect by insulin and
glucagon awaits further study.

Effects of Insulin and Glucagon on Fetal Heart

Guidotti and his co-workers pioneered studies of the
ontogeny of metabolic responses to insulin using embryonic
chick heart as a model. They showed that insulin stimulated glucose uptake at a later stage of development than the
incorporation of acetate into lipid, but before the uptake
of amino acids and their incorporation into protein. These
changes take place early, between the fifth and tenth day
in ovo.

This methodological approach has been extended more
recently using the fetal rat heart. The basal rate of
glucose uptake by isolated fetal cardiac muscle fell from
day 16 to 22 and remained constant after birth. The rate
of glucose incorporation into glycogen rose to a maximum on
day 18 and cardiac glycogen was at its highest concentration on day 21. Carbon dioxide production from glucose was
less than 2% and did not change with age. The addition of
a pharmacological concentration of insulin to the
incubation increased glucose uptake by a similar fraction
(approximately 70%) in 16, 18, 21 day fetuses and 1 day
newborn rats. By comparing the digestion of cardiac
glycogen with cold or hot potassium hydroxide, the glucose
uptake stimulated by insulin was shown to be incorporated
into a glycogen moiety that was turning over rapidly (30).

Abnormalities of pregnancy such as maternal alloxan
diabetes and fasting have also been studied (31). Fasting
for 16 h immediately prior to sacrifice at term caused
maternal plasma insulin and glucose concentrations to fall
and led to a drop in fetal plasma insulin, hepatic and
myocardial glycogen despite a rise in blood glucose

concentration. Maternal diabetes induced on day 18 caused maternal and fetal hyperglycemia, no change in fetal plasma insulin, a fall in fetal hepatic glycogen but a rise in fetal cardiac glycogen concentration. Fetal myocardial glucose uptake in vitro was unaffected by maternal diabetes. These results show how a change in circulating substrate concentration can alter tissue carbohydrate in the face of a constant circulating hormone concentration.

In the same model insulin was shown to increase amino acid transport and protein synthesis from the 16th day of gestation onwards. The effect was independent of the presence of glucose in the incubation medium, was insulin dose-related and appeared to act by both the alanine and leucine receptor pathways.

Glucagon acted on rat heart at a later stage of development than insulin and by a final common pathway shared with epinephrine which makes easier the analysis of the point at which glucagon failed to act. Fetal heart on day 16, 18 and 20 responded to epinephrine with an increase in adenylate cyclase activity and a fall in glycogen concentration but did not response to glucagon in the same way until 4 weeks postnatally (32). In the 19 day mouse fetus epinephrine and glucagon both caused a depletion of cardiac glycogen and an increase of atrial concentration rate (33). These results illustrate differences between organs, the heart and liver in the development of receptors to glucagon, and differences between related species, the rat and mouse in the time at which glucagon sensitivity appears in the heart.

Studies have been performed on human fetal heart homogenates of 5-17 weeks gestation (34). Adenylate cyclase could be activated by sodium fluoride at all ages studied, by epinephrine from 6-7 weeks and by glucagon from 8-9 weeks onwards.

Effects of Insulin on Fetal Muscle and Skeleton

Early experiments indicated that glucose uptake by the
fetal rat diaphragm _in_ _vitro_ at term was not affected by
insulin and that neither insulin nor insulin-antibody
influenced the accumulation of radioactive glucose by hind
limb muscle. Muscle sensitivity to insulin appeared
shortly after birth. More recently contrary results were
obtained (35). Insulin at a physiological concentration
(5 µU/ml) stimulated the uptake of radioactive
2-deoxyglucose, α-amino isobutyric acid or leucine by
isolated diaphragm from 20 day fetal to 14 day neonatal
rats. Methodological arguments were invoked to explain the
differences. The observation has physiological importance:
59% of total body glycogen is in rat fetal muscle on the
21st day compared with 50% on the 20th day (36) and if
this can be influenced by circulating insulin there could
be important implications for perinatal carbohydrate
homeostasis.

Despite much work on the development of metabolic
pathways in fetal rhesus monkey skeletal muscle, little
attention has been paid to the hormonal factors involved.
Term in the rhesus monkey is 165 days. Fetal thigh muscle
at 52 to 61% of term was used to study the effects of
insulin (10 µU/ml) or epinephrine (6 x 10^{-6}M) on carbohydrate metabolism _in_ _vitro_ (37). Insulin increased
glucose uptake and incorporation into glycogen, lactate
and carbon dioxide production. Epinephrine decreased
glucose uptake, glycogen content and carbon dioxide
production. The two hormones appeared to affect glycogen
metabolism as early as 85 days gestation via cyclic AMP
and similar enzyme systems to those that operate in the
adult.

After birth, longitudinal skeletal growth in mammals
is thought to be mediated by the somatomedins, which are
largely dependent on growth hormone (38). The control of
fetal skeletal growth may be different because in several
species growth has been shown to progress normally after

decapitation or hypophysectomy in utero and we have shown
that somatomedin activity is retained in the fetal rabbit
following decapitation in utero (39). This led to further
experiments in which the control of fetal somatomedin was
studied (40). One fetus in each of 12 litters was injected
with 1 unit of IZS insulin subcutaneously on day 27 of
gestation and a control fetus was injected with the same
volume of saline. The litter was delivered by caesarean
section on day 29 and each fetus identified. Plasma
somatomedin activity was determined by fetal rabbit
cartilage bioassay. Costal cartilage from individual
fetuses was incubated in medium containing $[^3H]$ thymidine
or $[^{35}S]$ sulphate as indicators of cell replication and
matrix synthesis respectively. Individual values for
somatomedin activity or cartilage isotope uptake were
ranked within a litter. In each case the rank in the litter of the insulin-injected fetus, but not the saline-
injected fetus, was significantly higher than the mean
rank of the litter. Insulin did not stimulate cartilage
metabolism in vitro. The results suggest that insulin
may stimulate fetal skeletal growth via the somatomedin
pathway.

The availability of human fetal cells that might be
representative of the lean body mass is limited but
fibroblasts and mononuclear leucocytes have been used to
study the biological effects of insulin. Monolayer
fibroblast cultures were prepared from fetuses of 78 to
127 days gestational age. Cultures in serum-free medium
with insulin (0.1 to 100 mU/ml) resulted in the stimulation
of glucose uptake, uridine incorporation into RNA and
leucine incorporation into protein which persisted for
60 min after removal of insulin (41). Competitive binding
studies suggested that the prolonged effect was due to
persistent interaction of insulin with its receptor site.
Monocytes harvested from the cord blood of normal newborns
bound five times as much insulin as those of healthy
adults (42). The difference was due to increased receptor
affinity and in the number of receptor sites per cell.

An inference from both studies is that insulin may have a biological effect in the lean body mass of the human fetus from early in prenatal life.

Effects of Insulin and Glucagon on Fetal Adipose Tissue

In comparison to the organs and tissues considered above there is a surprising paucity of information about the effects of insulin and glucagon on fetal and neonatal adipose tissue. This is partly because adipose development in the most intently studied species, the rat, takes place mainly after birth. In a major review of the development of adipose tissue (43) emphasis was placed on the physiological and metabolic differences of brown and white adipose tissue (BAT and WAT) and on the heterogeneity of WAT between the sexes or from different sites in the body. The major evidence for a lipogenic effect of insulin remains the clinical observations on IDM and infants suffering from neonatal diabetes mellitus.

The actions of glucagon on adipose tissue are complicated. BAT from newborn rats or rats older than 20 days released glycerol in vitro when exposed to glucagon but was unresponsive when taken from animals aged 2-20 days. Subcutaneous WAT became responsive to glucagon after day 18 but did not react to the hormone later in life. Ovarian and epididymal WAT showed an increasing responsiveness to glucagon with age however. The evidence indicates that glucagon does not have a primary role in the control of early neonatal lipolysis which is effected by catecholamine release, but may have an important secondary role by raising tissue glucose levels via glycogenolysis in WAT and/or hepatic gluconeogenesis (43).

SUMMARY

The metabolic effects of insulin and glucagon in prenatal life are reviewed in man and laboratory animals. The two hormones often have opposing actions and their net effect on the fetus is determined not only by the con-

centration of each hormone in the circulation but also by
the development of receptors on target cells. For example
in the rat there is an insulin/glucagon molar ratio of 10
late in gestation but insulin receptors have not developed
in certain tissues (e.g. skeletal muscle). By contrast in
man tissue sensitivity to insulin and glucagon appears
early in fetal life but there is not such an extreme molar
ratio in favour of either hormone.

Each hormone influences fetal growth indirectly via an
effect on the establishment of energy reserves. In normal
pregnancy insulin plays the dominant role by stimulating
the accumulation of carbohydrate and lipid particularly
in the liver and adipose tissue. If substrate delivery is
impaired as in maternal starvation, fetal hyperglucagonemia
may occur resulting in fetal glycogenolysis and the premature induction of gluconeogenesis. Glucagon deficiency in
the fetus does not appear to impede fetal growth.

Insulin also stimulates amino acid uptake and protein
synthesis by fetal cells, but the contribution made in this
way to either normal or abnormal fetal growth (infant of a
diabetic mother, transient neonatal diabetes mellitus) is
unknown.

ACKNOWLEDGEMENTS

Financial support from the Medical Research Council,
British Diabetic Association and the Wellcome Trust is
gratefully acknowledged.

REFERENCES

1. Adam, PAJ, K Teramo, N Raiha, D Gitlin and R Schwartz,
    Human fetal insulin metabolism early in gestation.
    Response to acute elevation of the fetal glucose
    concentration and placental transfer of human
    insulin-I-131. Diabetes 18:409-416,1969.

2. Moore, WMO, BS Ward and C Gordon. Human placental
    transfer of glucagon. Clin Sci Molecular Med
    46:125-129, 1974

3. Pregnancy Metabolism, Diabetes and the Fetus,
    Elliott, K and O'Connor, M (eds.), Amsterdam,
    Excerpta Medica, 1979

4. Carbohydrate Metabolism in Pregnancy and the Newborn
    1978, Sutherland, HW and Stowers, JM (eds.),
    Berlin, Springer-Verlag, 1979

5. Roversi, GD, M Gargiulo, U Nicolini, E Pedretti,
    A Marini, V Barbarani and P Peneff, A new approach
    to the treatment of diabetic pregnant women:
    479 cases (1963-1975). Am J Obstet Gynec 135: 567,
    1979

6. Spergel, G, LJ Levy and MG Goldner, Glucose intoler-
    ance in the progeny of rats treated with single
    subdiabetogenic dose of alloxan. Metabolism 20:
    401-413, 1971

7. Aerts, L and FA Van Assche, Is gestational diabetes an
    acquired condition? J Devel Physiol, in press

8. Milner, RDG, The growth and development of the
    endocrine pancreas, In: Scientific Foundations of
    Paediatrics, Davis, JA and Dobbing, J (eds.),
    London, William Heinemann Medical Books Ltd.
    2nd ed. 1980 in press.

9. Manns, JG and RP Brockman, The role of insulin in the
    synthesis of fetal glycogen. Canad J Physiol
    Pharmacol 47:917-921, 1969

10. Picon, L, F Bailly, A Kervran and M Rieutort,
    Hyperglycémie chez le foetus de rat injecté de
    serum anti-insuline. CR Acad Sci Paris 271:
    774-776, 1970

11. Rabain, F and L Picon, Effect of insulin on the
    materno-fetal transfer of glucose in the rat.
    Horm Metab Res 6:376-381, 1974

296

12. Girard, JR, GS Cuendet, EB Marliss, A Kervran,
    M Rieutort and R Assan, Fuels, hormones and
    liver metabolism at term and during the early
    postnatal period in the rat. J Clin Invest 52:
    3190-3200, 1973

13. Girard, JR, P Ferre, M Gilbert, A Kervran, R Assan
    and EM Marliss, Fetal metabolic response to
    maternal fasting in the rat. Am J Physiol 232:
    E456-E463, 1977

14. Sherwood, WG, GW Chance and DE Hill, A new syndrome
    of pancreatic agenesis. Pediat Res 8:360, 1974

15. Shelley, HJ, JM Bassett and RDG Milner, Control of
    carbohydrate metabolism in the fetus and newborn.
    Brit Med Bull 31:37-43, 1975

16. Chez, RA, DH Mintz, EO Horger and DL Hutchinson,
    Factors affecting the response to insulin in the
    normal subhuman pregnant primate. J Clin Invest
    49: 1517-1528, 1970

17. Reisner, SH, JV Aranda, E Colle, A Papageorgiou,
    D Schiff, CR Scriver and L Stern, The effect if
    intravenous glucagon on plasma amino acids in the
    newborn. Pediat Res 7:184-191, 1973

18. Vidnes, J and S Øyasaeter, Glucagon deficiency causing
    severe neonatal hypoglycemia in a patient with
    normal insulin secretion. Pediat Res 11:943-949,
    1977

19. Grajwer, LA, MA Sperling, J Sack and DA Fisher,
    Possible mechanisms and significance of neonatal
    surge in glucagon secretion studies in newborn
    lambs. Pediat Res 11:833-836, 1977

20. Sperling, MA, LA Grajwer, RD Leake and DA Fisher,
    Effects of somatostatin (SRIF) infusion on glucose
    homeostasis in newborn lambs: evidence of a
    significant role of glucagon. Pediat Res 11:
    962-967, 1977

21. Warnes, DM, RF Seamark and FJ Ballard, The appearance
    of gluconeogenesis at birth in sheep. Activation
    of the pathway associated with blood oxygenation.
    Biochem J 162:627-634, 1977

22. Hunter, DJS, Changes in blood glucose and liver
    carbohydrate after intrauterine injection of
    glucagon into foetal rats. J Endocrinol 45:
    367-374, 1969

23. Blazquez, E, B Rubalcava, R Montesano, L Orci and RH Unger, Development of insulin and glucagon binding and the adenylate cyclase response in liver membranes of the prenatal, postnatal and adult rat:evidence of glucagon "resistance". Endocrinology 98:1014-1023, 1976

24. Greengard,), Enzyme differentiation of human liver: comparison with rat model. Pediat Res 11: 669-677, 1977

25. Schwartz, AL, Influence of glucagon dibutyryl cyclic AMP and triamcinolone on the arginine synthetased system in perinatal rat liver. Biochem J 126: 89-98, 1972

26. Girard, JR, D Caquet, D Bal and I Guillet, Control of rat liver phosphorylase, glucose-6-phosphatase and phosphoenolpyruvate carboxykinase activities by insulin and glucagon during the perinatal period. Enzyme 15:272-285, 1973

27. Schwartz, AL and TW Rall, Hormonal regulation of incorporation of alanine U-14-C into glucose in human fetal liver explants. Effect of dibutyryl cyclic AMP, glucagon, insulin and triamcinolone. Diabetes 24:650-657, 1975

28. Schwartz, AL, NCR Raiha and TW Rall, Hormonal regulation of glycogen metabolism in human fetal liver. 1. Normal development and effects of dibutyryl cyclic AMP, glucagon and insulin in liver explants. Diabetes 24:1101-1112, 1975

29. Schwartz, AL, Hormonal regulation of amino acid accumulation in human fetal liver explants. Biochem Biophys Acta 363:276-289, 1974

30. Clark Jr, CM, Characterization of glucose metabolism in the isolated rat heart during fetal and early neonatal development. Diabetes 22:41-49, 1973

31. Vinicor, F, D Kohalmi and CM Clark Jr, Characterization of carbohydrate metabolism in the isolated fetal rat heart. Effects of fasting and alloxan diabetes. Diabetes 23:662-669, 1974

32. Clark Jr, CM, DO Allen and JF Clark, Appearances of responses to glucagon in cultured neonatal rat heart cells. Endocrinology 100:989-993, 1977

33. Wildenthal, K., DO Allen, J Karlsson, JR Wakeland and
    CM Clark Jr, Responsiveness to glucagon in fetal
    hearts - species variability and apparent
    disparities between changes in beating, adenylate
    cyclase activation and cyclic AMP concentration.
    J Clin Invest 57:551-558, 1976

34. Palmer, GC and WG Dail Jr, Appearance of hormone
    sensitive adenylate cyclase in developing human
    heart. Pediat Res 9:98-103, 1975

35. Fricke, R and CM Clark Jr, Augmentation of glucose
    and amino acid uptake by insulin in the developing
    rat diaphragm. Am J Physiol 224:117-121, 1973

36. De Meyer, R, P Gerard and G Verellen, Carbohydrate
    metabolism in the newborn rat, In: Metabolic
    Processes in the Foetus and Newborn Infant,
    Jonxis, JHP, Visser, HKA and Troelstra, JA (eds.),
    Leiden, H.E. Stenfert Kroese N.V. 281-291, 1970

37. Bocek, RM, MK Young and CH Beatty, Effect of insulin
    and epinephrine on the carbohydrate metabolism and
    adenylate cyclase activity of rhesus fetal muscle.
    Pediat Res 7:787-793, 1973

38. Van Wyk, JJ and LE Underwood, Relation between growth
    hormone and somatomedin. Ann Rev Med 26:427-441,
    1975

39. Hill, DJ, P Davidson and RDG Milner, Retention of
    plasma somatomedin activity in the foetal rabbit
    following decapitation in utero. J. Endocrinol,
    81:93-102, 1979

40. Hill, DJ and RDG Milner, Increased somatomedin
    and cartilage metabolic activity in rabbit fetuses
    injected with insulin in utero. Diabetologia
    1980 in press

41. Fujimoto, WY and RH Williams, Persistent biologic
    actions of insulin on cultured fetal human fibro-
    blast. In Vitro 13:268-274, 1977

42. Thorsson, AV and RL Hintz, Insulin receptors in the
    newborn: an increase in affinity and number.
    New Engl J Med 297:908-912, 1977

43. Hahn, P and M Novak, Development of brown and white
    adipose tissue. J Lipid Res 16:79-91, 1975

DISCUSSION

Qu.Salle : Has growth hormone a role in regulation of somatomedin activity in the fetus?

An.Milner : There in no clear answer to this question at
present but evidence currently available does not support
the idea that fetal somatomedin is dependent on fetal
growth hormone (Hill, D.J., Davidson, P. and Milner, R.D.G.
1979, J. Endocrinol. 81, 93-102).

Com.Teller : Glibendamide treated members of a rat litter
revealed higher concentrations of insulin and somatomedin
(porcine cartilage assay) in blood than their untreated
litter mates (Heinze et al., Diab.Fed.Meeting, Vienna,
1979).

Qu.Helge : How long does it take until you can demonstrate
these effects of an insulin injection in the rabbit fetus
at day 27 of pregnancy? Can you find somatomedin activity
increased already after 6 to 12 hours? Insulin injections
at other stages of fetal development, do they have similar
effects?

An.Milner : We have not performed experiments in which fetuses have been harvested earlier or have been injected
with insulin at other stages of pregnancy.

PLASMA CORTICOSTEROIDS IN NORMAL, PREMATURE AND "SMALL-FOR-DATES" NEWBORN INFANTS THROUGHOUT THE NEONATAL PERIOD

W. Rokicki and J. Bertrand

Plasma content of cortisol (F), cortisone (E), corticosterone (B), 11-dehydrocorticosterone (A) and 11-deoxycortisol (S) has been determined in the peripheral venous blood of 77 full-term, 93 premature and 49 small-for-dates newborns from birth up to the third month of life.

The infants studied were born in the Clinic of Pathological Pregnancy (Institute of Obstetrics and Gynecology, Silesian Medical Academy, Poland), and in several other hospitals of Upper Silesia (Poland) as well as in the Clinique Saint-Charles, Hospital of the University of Montpellier (France).

The newborns were divided into the following groups : 0-6 hours (n = 17), 7-12 hours (n = 14), and 13-24 hours (n = 15) after birth, second day of life (n = 31), 3rd-5th day (n = 42), 6th-12th day (n = 41), 13th-20th day (n = 39), second and third month (n = 18).

The plasma organic extracts were chromatographed on Sephadex LH-20 columns. Separate chromatographic fractions were taken for the determination of Kendall' A, B, S, E and F compounds by a radioimmunoassay using a non specific antibody, raised in rabbits against cortisone-21-BSA, and kindly donated by Dr. Bidlingmaier (Children's Hospital, München) (2).

Part of the results are shown in figures 1-4. The evolution of total plasma corticosteroid (sum of A, B, S, E and F compounds) in the three groups can be compared in figure 1. Plasma total corticosteroids were significantly more elevated in prematures between 6 hours and 4 days of life (p< 0.01) than in the other groups, and prepubertal children (mean value 10.5 µg/dl for the latter (2)). However the three groups followed roughly the same pattern, with a progressive decrease of plasma corticosteroid concentration from birth to the end of the second month.

Plasma levels of cortisol in full-term newborns decreased rapidly during the first 24 hours of life and then remained at a low plateau from day 2 to day 30 with values significantly lower than those observed in prepubertal children (day 2 = 2.93 ± 1.92 (DS) μ g/dl ; day 3-5 = 2.29 ± 0.83 (DS) ng/dl ; day 6-12 = 4.87 ± 1.92 (DS) μg/dl ; day 13-30 = 4.38 ± 2.47 (DS) ng/dl ; prepubertal children = 8.9 ± 1.7 (DS) μ g/dl) (1, 2) (Fig. 2).

- In all groups plasma levels of cortisone were much higher than those in prepubertal children (1.35 ± 0.3 μg/dl (2)) and a progressive decrease from birth to the third month of life was observed (figure 3).

- Consequently, during this period the cortisone over cortisol ratios (E/F) remained higher in newborns than in children (0.2 (0.15-0.25) (1, 2) (fig. 4). However the pattern was quite different among each newborn group : full-term and small-for-dates E/F ratios varying between 2 and 3 which is higher than in premature newborns.

These results allow the following statements to be made :

- During the first 5 days of life the increase of total plasma corticosteroids is higher in premature than in full-term or small-for-dates infants.

- An increase of 11 oxidised biologically inactive forms over 11 β - hydroxylated biologically active forms is observed in all groups with an E/F ratio five (in prematures) to ten (in full-terms) times higher in infants than in children.

- Thus during the first 5 days of life, cortisol levels in full-term and small-for-dates infants are much lower than in prepubertal children (3). In contrast plasma cortisol is higher in prematures than in prepubertal children during this period (1).

REFERENCES

1. Klein, G, M Baden, and C Giroud, Quantitative measurement and significance of five plasma corticosteroids during the perinatal period. J Clin Endocrinol Metab 36:944-950, 1973.

2. Sippel, WG, F Bidlingmaier, H Becker, T Bronig, H Dörr, H Hahn, W Golder, G Hollmann, and D Knorr, Simultaneous radioimmunoassay of plasma aldosterone, corticosterone, 11-deoxycorticosterone, progesterone, 17-hydroxyprogesterone, 11-deoxycortisol, cortisol and corticosterone. J Steroid Biochem 9:63-74, 1978.

3. Sippel, W, H Becker, H Versmold, F Bidlingmaier, and D Knorr, Longitudinal studies of plasma aldosterone, corticosterone, deoxycorticosterone, progesterone, 17-hydroxyprogesterone, cortisol and cortisone determined simultaneously in mother and child at birth and during the early neonatal period. I. Spontaneous delivery. J Clin Endocrinol Metab 46:971-985, 1978.

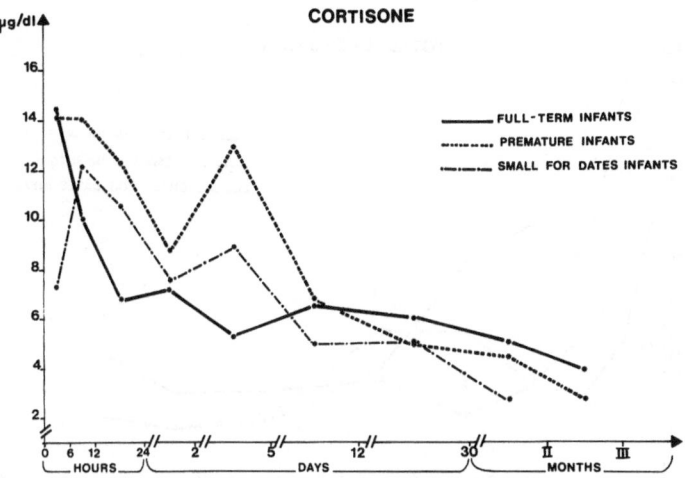

Figure 3 : Pattern of plasma cortisone levels in full-term premature and small-for-dates infants from birth to third month of life. Control values in children = 1.35 ± 0.3 (DS) µg/dl (2).

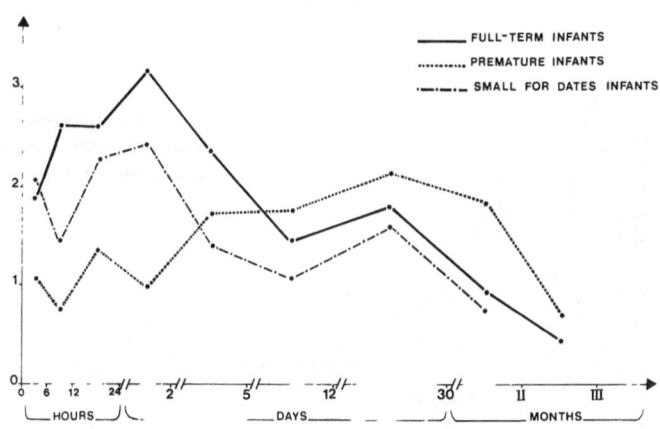

Figure 4 : Pattern of cortisone (E) over cortisol (F) ratios in plasma of full-term, premature and small-for-dates infants from birth to third month of life. Control values in children = 0.15-0.25 (1, 2).

304

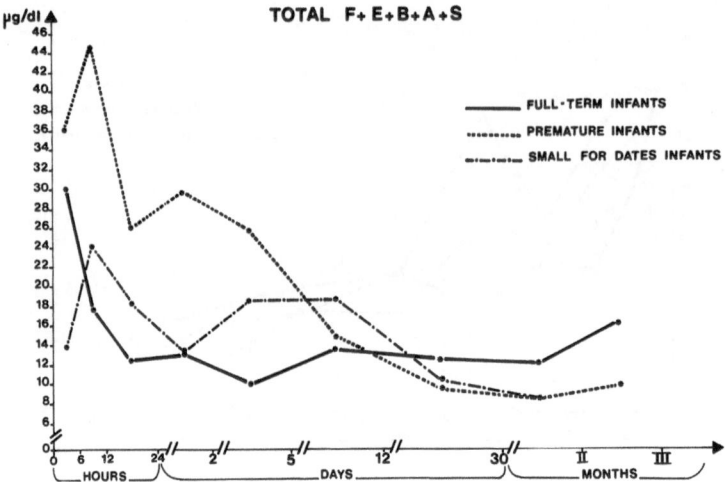

Figure 1 : Pattern of total plasma glucocorticoid (F + E + A + B + S compounds) levels in full-term, premature and small-for-dates infants from birth to third month of life. Control value in children = 10.5 µg/dl (2).

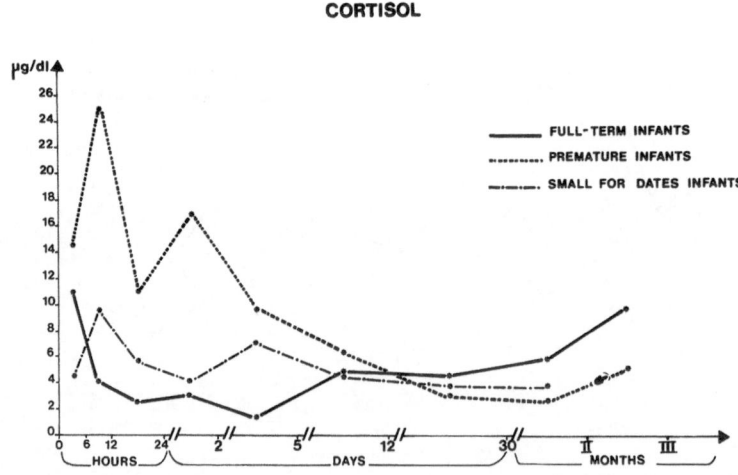

Figure 2 : Pattern of plasma cortisol levels in full-term, premature and small-for-dates infants from birth to third month of life. Control values in children = 8.9 ± 1.7 (DS) µ g/dl (2).

DISCUSSION

Qu.Sodoyez : What is the level free (diffusible) control in each group? Are the variations of F reflecting transcortex levels?

An.Bertrand : Free fraction of corticosteroids has been already measured by different authors : for more details see : Hadjian,A.J. et al., Pediatr.Research,1975,9,40.

Qu.Sodoyez : What is the level of progesterone (inhibition of control action on the hypothalamus) in each group?

An.Bertrand : Plasma level of progesterone has not been measured in this study. See Sippell et al.,Am.J.Obstet. Gynecol.,1979,135,530.

Qu.Bossi : You have presented data on basal steroid levels. Did you also find differences in reactivity of the adrenals upon stimulation between prematures, light for gestational age and full term age babies?

An.Bertrand : There is a general adreno-cortical hyperresponsiveness to corticotropin stimulation during the first month of life, whatever newborn group is. For more details see : Forest, M.G. et al.,J.Clin.Endocrinol.Metab.,1978, 47/5,931.

Qu.Helge : Dr.Korth-Schütz in Berlin has found similar differences between full-term and premature newborns also for DHEA, and there seems to be a correlation with the slow involution rate of the fetal adrenal in the premature. Therefore the immaturity and insensitivity might not be at the hypothalamo-hypophyseal level, as you suggested, but at the adrenal.

An.Bertrand : This slow involution rate of fetal adrenal cortex does not mean a decrease in adrenal responsiveness to ACTH stimulation since in premature plasma cortisol levels are higher than in children during the first five days of life.

Qu. Teller : What are the dynamics of corticosterone
in the 3 different groups of neonates ?
An. Bertrand : The course of plasma corticosterone levels
has been also studied in the 3 groups of newborns. It is
the same as cortisol.

SOMATOMEDIN A ACTIVITY
IN NEWBORNS

A. Ruiton-Ugliengo, B. Salle, G. Putet, J. Bertrand, G. Tell

The present study concerns S.M.A. A activity with the following aims :
- to compare arterial and venous cord blood in prematures,
full term newborns and small for gestational age infants
(S.G.A.).
- To follow the hormone evolution during the first days of
life and compare the results of each group with each other.

MATERIAL AND METHODS

Clinical data are shown in Table I. Prematures and S.G.A.
infants were perfused with 10 % glucose, 80 ml/kg/day but
full terms were not. Feeding was started in full term
newborns at six hours and the others at twelve hours of age.
At birth, mother and cord blood samples (artery and vein)
were collected; later on, blood samples were taken at two
hours, two days and six days. Blood was collected in a solution of sodium citrate and was centrifuged immediately;
plasma was stored at -25°C. Until assay, somatomedin A
activity was assayed with the technic of Hall (1) modified
by Audhya and Gibson (2). It is a biologic assay using
incorporation of sulfate by chicken embryo cartilage incubated for twelve days.
The radioactive sulfate incorporation in cartilage is measured both with a standard plasma and with a sample plasma.
As the cartilage used cannot be perfectly indentical, its
weight cannot be taken into account when calculating the

result; therefore this method is difficult because of the
weighing. Gibson (2) has introduced a modification of this
method by suppressing the weighing and using a standard
range of sulfate chondroitine colored by alcian blue and
measured photometerically. Alcian blue staining of cartilage after papain is followed by the measurement of optic
density. This permits to calculate the weight of cartilage
in μg of chondroitine sulfate. On the aliquot part of the
sample, S_{35} beta radioactivity which is bound to the cartilage, is counted after calculation; the number of cpm is
transformed into μg of chondroitine sulfate. This value is
proportional to the S.M.A. A activity in the plasma sample.

RESULTS

Somatomedin A activity of normal adults plasma pool is
defined as being equal to 1 unit for 1 ml. In children
with growth hormone retardation, S.M.A. A is between 0.6
and 0.7.
a) Arterial and venous cord blood : in full term newborns,
plasma S.M.A. A levels in venous blood is significantly
higher than in arterial blood ($p < 0.0017$). In premature
and S.G.A. infants there was no significant difference between arterial and venous cord blood ($p > 0.05$) (figure 1.).

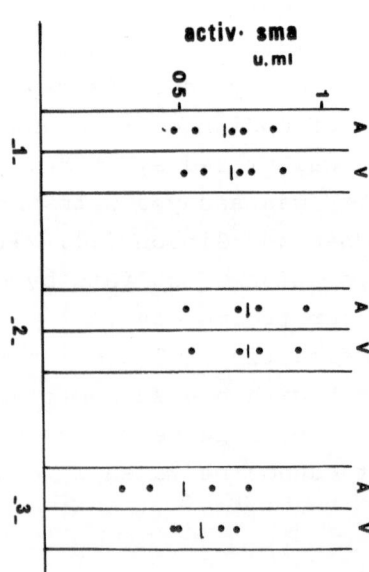

Figure 1 : Somatomedin A activity in arterial and
           venous cord blood
        1 - Full term newborns
        2 - Premature infants
        3 - Small for gestational age infants

b) If we consider each group, there was a significant difference in plasma S.M.A. levels between cord blood and blood of two days old children in full term and in premature newborns (p< 0.05) but not in S.G.A. infants (Table II)
c) In each group, there was no correlation between mean cord plasma S.M.A. levels and body or placenta weight.
d) There was no correlation in plasma S.M.A. levels between mother and cord blood in either of the group.

	NUMBER	G.A. (weeks)	B.W. (g)	HEIGHT (cm)	H.C. (cm)
Full-term Newborns	5				
Mean		39.8	3560	50.8	35.6
±		±	±	±	±
SD		1.1	547	2.6	1.3
Prematures	8				
Mean		35.37	2172	45.5	31.06
±		±	±	±	±
SD		0.9	292	1.5	1.8
Small for Dates	8				
Mean		37.8	1890	43.25	30.37
±		±	±	±	±
SD		0.8	171	1.3	1.2

TABLE I

CLINICAL DATA

	MOTHER BLOOD	CORD BLOOD	2 H	DAY 2	DAY 6
FULL TERM	0.76 ± 0.09 (n = 5)	0.66 ± 0.14 (n = 5)	0.76 ± 0.08 (n = 5)	0.97 ± 0.14 (n = 5)	0.93 ± 0.14 (n = 2)
PREMATURES	0.72 ± 0.08 (n = 4)	0.53 ± 0.16 (n = 8)	0.69 ± 0.19 (n = 8)	0.72 ± 0.09 (n = 8)	0.68 ± 0.14 (n = 6)
S.G.A.	0.71 ± 0.09 (n = 3)	0.61 ± 0.17 (n = 8)	0.69 ± 0.13 (n = 7)	0.70 ± 0.2 (n = 7)	0.80 ± 0.15 (n = 4)

TABLE II

Somatomedin A activity (mean and SD) in mother , cord blood, 2 hours, 2 days and 6 days (sequential study)

REFERENCES

1. Hall, K, Human somatomedin. Acta Endocr suppl. 163:1, 1972
2. Audhya, T.K. and K.D. Gibson, Serum inorganic sulfate and apparent somatomedin activity in assay using chick embryo cartilage. Endocrinology 6:1614, 1974
3. Hall, K, Quantitative determination of sulfatation factor activity in human serum. Acta Endocr 2:338, 1970
4. Kastrup, K.W., M.J. Andersen and P. Lebech, Somatomedin in newborns and the relationship to human chrionic somatotropin and fetal growth. Acta Pediatr Scand 6:757, 1978
5. Hintz, R.L., J.M. Seeds, R.E. Johnson-Baugh, Somatomedin and growth hormone in the newborn. Amer J Dis Child 11:1249, 1977
6. Tato, L., M.V.L. du Caju, C. Prevot and R. Rappaport, Early variations of plasma somatomedin activity in the newborn. J Clin Endocrinol Metab 3:534, 1975
7. Gluckman, P.D., and M.W. Brinsmead, Somatomedin in cord blood. Relationship to gestational age and birth size. J Clin Endocrinol Metab 6:1378, 1976
8. Svan, H., K. Hall, M. Ritzen, K. Takano and A. Skottner, Somatomedin A and B in serum from neonates their mothers and cord blood. Acta Endocrinol 3:636, 1977.

DISCUSSION

In the literature, different authors have used different techniques of assay (bioassay using different cartilage chicken or porc, assay using radioreceptor, etc...) and this makes the comparison of these results extremely difficult (3, 4, 5, 6, 7, 8).

Hall et al. (3) have studied somatomedin A activity by radioreceptor (in cord and mother's blood) and found no correlation in the results; they concluded that the somatomedin was produced in the foetal-placental unit and they confirmed this fact by a high activity of S.M.A. in placental extracts. Kastrup (4) have measured S.M.A. A activity in maternal blood during the last trimester and at birth in mother's and cord blood of full term babies. They found a low activity in the mothers and newborns at birth but a correlation between mother's somatomedin levels and birth weight. Artos confirmed this data by studying the plasma S.M.A. in mothers and infants at birth.

In our experience, the mean plasma level of S.M.A. A in cord blood was low in each group of infants. These results

agree with those of Hall, Kastrup and Hintz (5). After
birth, we have demonstrated a rapid increase of S.M.A. A
activity at 2 days, but this increase was more noticeable
in full term newborns. In S.G.A. infants, there was no
significant increase and the levels remained low (from
0.61 to 0.7 U/ml) and this result disagrees with those
of Tato and Rappaport (6). Mean plasma S.M.A. A level is
significantly higher in venous cord blood than in arterial
cord blood, in full term newborns but not in prematures or
S.G.A. infants. This suggests that the participation of
placenta in the synthesis of this peptide is more important
in full term then in low birth weight infants.

SESSION IV
PRACTICAL IMPLICATIONS AND
GENERAL DISCUSSION

EFFECT OF MATERNAL DIABETES ON GLUCOSE REGULATION IN THE NEWBORN

F.A. van Assche, L. Aerts, F. De Prins

INTRODUCTION

The infant of the diabetic mother is mainly oversized;
these macrosomic infants are born to hyperglycemic poorly
controlled non-ketotic diabetic mothers. Underweight
infants can be born to diabetic mothers with longstanding
diabetes and with vascular complications. In the overweight infants hypoglycaemia is frequently observed due to
the hyperinsulinism.

The aim of this presentation is to review, in the spirit of
the symposium, which antenatal factors are responsible for
the hyperinsulinism in the newborn of the diabetic mother.
We will concentrate on the morphologic and functional changes of the fetal and neonatal pancreatic B cells.

The changes in the fetal and neonatal B cells indicate
hyperstimulation and it seems very important to ascertain
whether or not these changes are reversible in later life.

1. The endocrine pancreas in the fetus and in the newborn
In the second half of fetal life and in the neonatal period
the islets of Langerhans are of the mantle type. The B
cells are the most numerous cells and are located in the
centre of the islets; the non-B cells are situated at the
periphery (mantle) of the islets. The non-B cells consist
of at least 6 different cell-types, which can be recognised
at the ultrastructural level. With immuno-cytochemical
techniques we have found cells containing the following

hormones : insulin, glucagon, somatostatin and human pancreatic polypeptide (HPP) (Van Assche et al., 1976; Van
Assche and Aerts, 1979). It has been shown that the B
cells contain insulin, the A cells glucagon and the D cells
somatostatin (Van Assche et al., 1976). The PP cells (or
the type V cells) probably contain human pancreatic polypeptide (Larsson et al., 1976, Van Assche et al., 1976).

2. The infant of the diabetic mother
One explanation for fetal and neonatal macrosomia in maternal diabetes is Pedersen's hypothesis (Pedersen, 1954).
This hypothesis suggests that fetal macrosomia is due to
hyperglycaemia of the pregnant mother. An accelerated placental tranfer of glucose is supposed to produce hyperglycaemia in the fetus, and this in turn causes pancreatic B
cell hyperplasia and fetal hyperinsulinaemia. The combination of hyperglycaemia and hyperinsulinaemia results in
accelerated de novo lipogenesis from glucose in the fetal
adipose tissue. Szabo and Szabo (1974) advanced an alternative hypothesis to explain the accelerated intra-uterine
adipogenesis. From the plasma of diabetic pregnant women,
free fatty acids (FFA) may be transferred across the placenta in increased quantities; these transferred serum
FFA's are subsequently taken up and esterified by the
adipocytes of the fetus. It is possible that maternal
hyperglycaemia and an increased level of circulating FFA
are both involved in fetal macrosomia (Van Assche and Aerts,
1979; Freinkel, 1979).
The abnormal intra-uterine milieu in diabetes and pregnancy
induces changes in the fetal endocrine pancreas. An
increased percentage of endocrine tissue and hyperplasia
of the B cells are well known characteristics in fetuses
and newborns of diabetic mothers (see review, Van Assche
and Aerts, 1979). Islet hypertrophy is due mainly to
hyperplasia of the B cells. The other cell types do not
show striking changes. Infants of diabetic mothers show
also an increased insulin level in the cord blood, as well
as an increased insulin content in the total pancreatic
tissue and in the microdissected islets (Steinke and

317

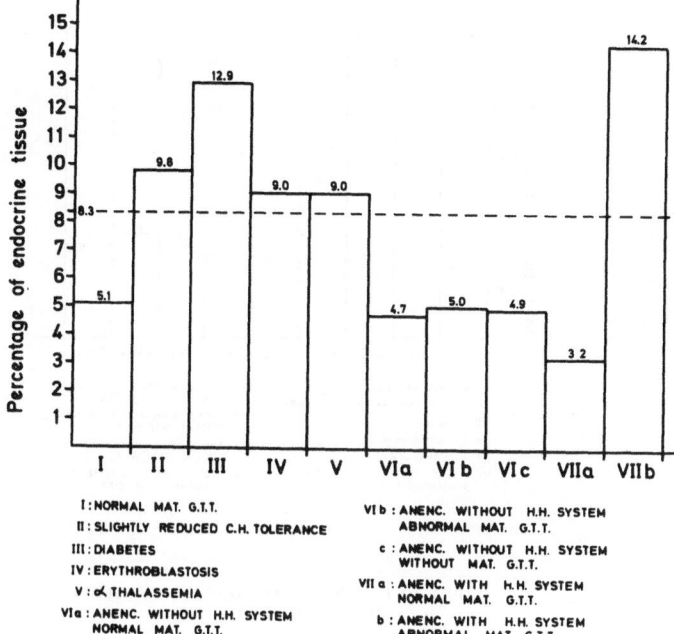

Figure 1

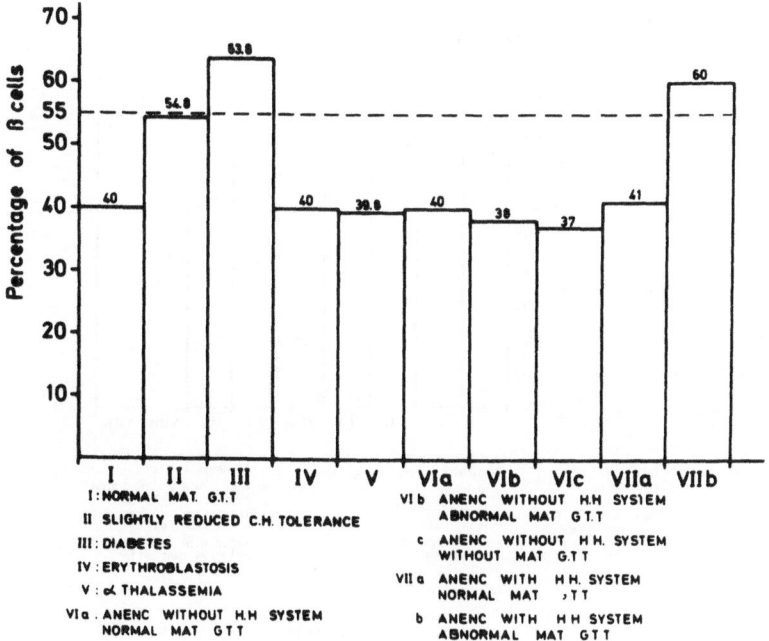

Figure 2

318

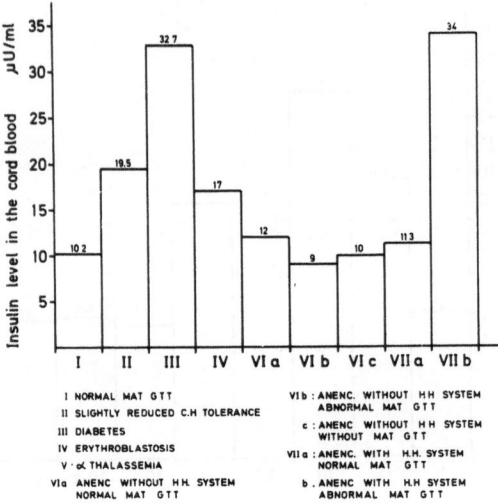

Figure 3

INSULIN CONTENT OF MICRODISSECTED ISLETS
IN THE DIFFERENT GROUPS

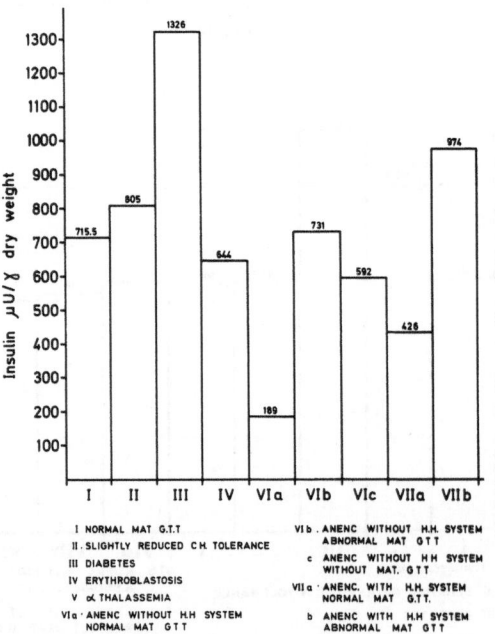

Figure 4

Driscoll, 1965; Van Assche, 1970). Fig. 1 clearly shows
that the percentage of endocrine tissue is more than twice
as high in fetuses and new-borns of diabetic mothers
(Group III) than in normals (Group I).
It is also clear that the amount of endocrine tissue is
increased in infants affected by Rhesus anti-D isoimmunisation (Group IV) and in infants born to mothers with a-
Thalassaemia (Group V). Anencephalics without a functional
hypothalamo-hypophyseal (H-H) system born to diabetic
mothers (Group VIb) have a normal amount of endocrine tissue;
however, anencephalics with a functional hypothalamo-hypophyseal system born to diabetic mothers (Group VIIb) have
an increased amount of endocrine tissue.
Fig. 2 gives data on the percentage of B cells in the endocrine pancreas. There is a net increase in the percentage
of B cells in infants of diabetic mothers (Groups II & III).
However, B cell hyperplasia is not present in erythroblastosis nor in infants of mothers with d Thalassaemia.It is
also clear that in anencephalics without a functional HH
system born to diabetic mothers (Group VIb) no B cell
hyperplasia is present, which, however, is found in anencephalics with a functional HH system, born to diabetic
mothers (Group VIIb). The insulin level in the cord blood
is increased in infants of diabetic mothers (Group II & III)
and also in erythroblastotic infants (Group VI). In anencephalics born to diabetic mothers an increased insulin
level is found only in the presence of a functional HH
system (Fig. 3).
The insulin content in the microdissected islets is
increased in infants of diabetic mothers (Groups II & III),
but not in erythroblastosis (Group IV). In the anencephalics born to diabetic mothers an increased insulin content
is also found, but only when a functional hypothalamo-
hypophyseal system is present (Group VII b).
From these observations it is clear that the volume of
insulin producing B cells is increased in infants of diabetic mothers. This B cell hyperplasia is caused by hyperglycaemia. In infants affected by anti-D isoimmunisation,
the proportion of B cells in the islets is not increased,

but the total endocrine mass and the insulin level in the
cord blood, are increased. This explains why hypoglycaemia
is a frequent observation in neonates of diabetic mothers
and in neonates with erythroblastosis. It is interesting,
but so far unexplained, that an intact functional HH system
is necessary for the genesis of B cell hyperplasia.
Stowers et al (1980) have postulated that in infants of
diabetic mothers secretion of pancreatic glucagon is
triggered by the hypoglycaemia in the neonatal period. Glucagon stimulates the release of calcitonin, which inhibits
calcium release from bone and favours hypocalcaemia.
The oversized hyperinsulinic infant of the diabetic mother
is well documented: little information is available on the
endocrine pancreas in the intra-uterine growth retarded
fetus of the diabetic mother. Not enough data are available
to know whether or not these fetuses have hyperinsulinism.
It is an important question in view of the role of fetal
insulin as a growth promoting factor (Liggings, 1974; Van
Assche et al., 1977).
In 3 cases of growth retarded fetuses from diabetic mothers
we found no increased insulin level in the cord blood (Van
Assche et al., unpublished observation).

3. Long-term effect
It is evident that the endocrine pancreas of fetuses and
newborns of diabetic mothers show morphologic and functional features of hyperinsulinism. The question now arises
whether these changes are reversible after birth. It is
possible that the B cells are over-stimulated during intra-
uterine life and could sustain a permanent defect.
Hultquist and Olding (1975) have shown that in infants of
diabetic mothers there is increased fibrosis in the islets
of Langerhans 2 weeks after birth. It has also been shown
that the incidence of juvenile diabetes is 20 times higher
in children of diabetic mothers than in the control popula-
tion (Farguhar, 1969). We have tried to find an answer to
this problem by using an experimental model of diabetes and
pregnancy in the rat. These are striking similarities in
the morphologic and functional features between the endo-

crine pancreas of the rat fetus in experimental diabetes
and the endocrine pancreas of the human fetus of diabetic
mothers (Aerts and Van Assche, 1977).

Aerts and Van Assche (1979) have shown that even mild experimental diabetes in the rat (first generation) causes persistent changes in the endocrine pancreas of their offspring
(second generation) which are not perceptible in basal
conditions, but become apparent in situations stressing B
cell activity, such as an intravenous glucose load or
pregnancy. This inadequacy of B cell compensation in the
second generation offspring of diabetic rats may be interpreted as a possible cause of gestational diabetes, which
is even manifest in the fetuses of the third generation.

4. The clinical consequence of hyperinsulinism in the infant
   of the diabetic mother

The fetus responds to a glucose challenge by increasing the
proportion of B cells and as a consequence there is increased
insulin synthesis and secretion. Hyperinsulinism during
fetal life will inhibit the third trimester increased production of pulmonary lecithin (Smith et al., 1975) .
Tchobroutsky (1979) has demonstrated that the lecithin-
sphingomyelin ratio in well controlled diabetic pregnancies
is not different from the normal. Infants of diabetic
mothers have an increased risk of symptomatic hypoglycaemia
after birth and have a greater developing obesity and diabetes in later life, Farguhar, 1969; Shar and Farguhar,
1975).

If the human B cell is capable only of a limited number of
cell devidions, as has been shown in the rat (Logotheto-
polus, 1972; Aerts and Van Assche, 1979) B cell hyperplasia
during intra-uterine life could jeopardise the regenerative
potential in post-natal life.

All these findings endorse the clinical practice of optimal
diabetic control during pregnancy.

Infants born to diabetic mothers whose blood sugar levels
have been carefully controlled with insulin throughout
pregnancy have less B cell hyperplasia or none at all
(Van Assche, 1970; Hultquist, 1971). Persson (1974) has

demonstrated also that adequate metabolic control of the
diabetic pregnancy reduces the neonatal complications.
For the mother, pregnancy modifies maternal glucose homeostasis. Failure to respond to the increased requirements
of pregnancy, as is the case in diabetes, may have consequences for the endocrine pancreas in later life. O'Sullivan
et al (1971) have shown that gestational diabetics not
treated with insulin during pregnancy may show further
deterioration of their endocrine pancreatic function in
later life. This long-term deterioration is seldon seen in
the insulin treated mothers.

LIST OF LEGENDS

Fig. 1 : The volume of endocrine tissue in normal infants
and in infants from diabetic mothers and in other
neonatal abnormal conditions.

Fig. 2 : The proportion of B cells in normal infants and
in infants from diabetic mothers and in other
neonatal abnormal conditions.

Fig. 3 : The insulin in the cord blood in the previously
mentioned neonatal conditions.

Fig. 4 : The insulin content in the islets in the previously
mentioned neonatal condition.

1. AERTS, L, and VAN ASSCHE, FA, The rat foetal endocrine
   pancreas in experimental diabetes. J. Endocr. 73 :
   339-346, 1977.
2. AERTS, L, and VAN ASSCHE, FA, Is gestational diabetes
   an acquired condition ? Journal of developmental
   physiology (in press)
3. FARGUHAR, JW, Prognosis for babies born to diabetic
   mothers in Edinburgh. Archs. Dis. childh. 44 : 36-40,
   1969.
4. FREINKEL, N, PHELPS, RL, and METZGER, BE, Intermediary
   metabolism during normal pregnancy.
   In carbohydrate metabolism in Pregnancy and the
   newborn 1978, pp. 1-31. Ed. HW Sutherland and J
   Stowers. Berlin ; Springer-Verlag, 1979.
5. HULTQUIST, GT, Morphologiy of the endocrine organs in
   infants of diabetic mothers. In Diabetes Mellitus
   pp. 686-694. Ed. RR Rodriguez and J. Valance.
   Amsterdam : Excerpta Medica, 1971.
6. HULTQUIST, GT, and OLDING, L, Pancreatic islet fibrosis
   in young infants of diabetid mothers. Lancet 11 :
   1015-1016, 1975.
7. LARSSON, LI, SUNDLER, F, and HAKANSON, R, Pancreatic
   polypeptide. A postulated new hormone : identifica-
   tion of its cellular storage site by light and
   electron microscopic immunocytochemistry. Diabetologia 12 : 221-226, 1976.
8. LIGGINS, GK, The influence of the fetal hypothalamus
   and pituitary on growth. In size at birth. pp 165-
   183. Ciba foundation symposium 27 Ed. K. Elliot
   and J. Knight, London : Ciba foundation, 1974.
9. LOGOTHETHOPOLUS, J, Islet cell regeneration and neogenesis. In Handbook of physiology; endocrinology;
   endocrine pancreas, pp 67-76 Ed. DF Steiner and
   N. Freinkel. Baltimore : Williams and Wilkins, 1972.
10. O'SULLIVAN, JB, CHARLES, D, and DANDROW, RV, Treatment
    of verified prediabetes in pregnancy. Journal of
    Reproductive Medicine, 97 : 45-48, 1971.

11. PEDERSEN, J, Weight and lenght at birth of infants of
    diabetic mothers. Acta endocr. co   16 : 330-342
    1954.

12. PERSSON, B, Assessment of metabolic control in diabetic
    pregnancy. In size at Birth pp 247-273. Ciba foun-
    dation Symposium 27. Ed. K. Eliott and J. Knight.
    London : Ciba foundation, 1974.

13. SHAH, MPK, and FARGUHAR, JW, Children of diabetic
    mothers ; Subsequent weight. In Early Diabetes in
    Early Life pp 587. Ed. Camerini-Davalos and H.S.
    Cole. New York : Academic Press, 1975.

14. SMITH, BT, GIROUD, CPJ, ROBERT, M, and AVERY, ME,
    Insulin antagonism of cortisol action on Lecithin
    synthesis by cultured fetal lung cells. Journal of
    Pediatrics, 87 : 953-955, 1975.

15. STEINKE, J, and DRISCOLL, GG, The extractable insulin
    content of pancreas from fetuses and infants of
    diabetic and control mothers. Diabetes 14 : 573-578,
    1965.

16. STOWERS, JM, HEDING, LG, FISHER, PM, TREHARNE, IAL,
    ROSS, IS, SUTHERLAND, HW, BEWSHER, PD, RUSSELL, G, and
    PRICE, HV, The relationship between glucagon and
    Hypocalcaemia in infants of diabetic mothers. In
    Carbohydrate metabolism in pregnancy and the new-
    born 1978. pp 152-162. Ed. HW Sutherland and J.
    Stowers. Berlin : Springer-Verlag, 1979.

17. SZABO, AJ, and SZABO, O, Placental free patty acid
    transfer and fetal adipose tissue development : an
    explanation of fetal adiposity in infants of diabetic
    mothers. Lancet ii : 498-499, 1974.

18. TCHOBROUTSKY, C, Fetal assessment in diabetic pregnancy
    using non-steroid assessment. In Carbohydrate meta-
    bolism in pregnancy and the newborn pp 304-333. Ed.
    HW Sutherland and J. Stowers. Berlin: Springer-
    Verlag, 1979.

19. VAN ASSCHE, FA, The foetal endocrine pancreas. A
    quantitative morphologic approach; thesis, Leuven,
    1970.

20. VAN ASSCHE, FA, GEPTS, W, and AERTS, L, The fetal
    endocrine pancreas in diabetes (human). Diabetologia
    12 : 423, 1976.

21. VAN ASSCHE, FA, DE PRINS, F, AERTS, L, and VERJANS, M,
    The endocrine pancreas in small for dates infants.
    Brit. J. Obst. Gynaec., 84 : 751-753, 1977.

22. VAN ASSCHE, FA, and AERTS, L, The fetal endocrine
    pancreas. Contr. Gynec. Obstet., 5 : 44-57, 1979.

DISCUSSION

<u>Qu.Teller</u> : You did not specifically mention the number and
size of the A-cells in the infant of the diabetic mother.
What happens to them? Do they increase in number and hypertrophy as well as the B-cells?
<u>An.Van Assche</u> : There is no proportional increase in the
number of A cells, but since the total amount of endocrine
tissue is increased, there is an absolute increase in the
amount of A cells.

<u>Qu.Eggermont</u> : Eosinophilic infiltration of the pancreatic
islets from infants of diabetic mothers has been described.
Is this finding frequent and what does it mean?
<u>An.Van Assche</u> : We found eosinophic infiltration in about
30 % of the infants of diabetic mothers, we have studied.
It is not quite clear what this finding means, but it is
important that it is found also in cases where the mother
has not been treated with insulin.

<u>Qu.Verellen</u> : What could be the mechanism of the hypertrophy
of endocrine tissue in erythroblastosis foetalis?
<u>An.Van Assche</u> : We have no good explanation of the increased
amount of endocrine tissue without proportional B cell hyper
plasia.

<u>Qu.Helge</u> : You mentioned that in erythroblastosis the num-
ber of islets per pancreas is increased, but the number of
B-cells and insulin content per islet are not increased
above that of the controls. Doesn't this mean that total
number of B-cells per newborn and total insulin production
nevertheless,are higher than in a normal newborn? Couldn't
they bear a similar risk as children of diabetic mothers
with regard to disturbances of glucose tolerance in later
life?
<u>An.Van Assche</u> : There is islet hypertrophy, but without an
increased proportion of B cells. However, the total amount
of B cells is increased, which explains the hyperinsulinism
in these newborns.

Qu.Girard : Have you any explanation for the appearance of gestational diabetes in pregnant rats of second generation? Does it result from the incapacity of maternal pancreas to override the gestational insulin resistance or from an increased peripheral insulin resistance?

An.Van Assche : I think it is a combination of both factors. There is during pregnancy a stress for the pancreatic B cells. The pregnant rats of the second generation are less able to adapt to the increased requirements.

Qu.De Meyer : Do you have any information about erythro-blastosis associated with anencephaly?

An.Van Assche : Unfortunately no.

Qu.Shelley : Do you have any information about the pituitary factor necessary for pancreatic hyperplasia in the fetus of the diabetic mother? Is growth hormone or ACTH implicated?

An.Van Assche :It is possible that growth hormone ACTH and cortisol are involved. The oestrogens are also low in anencephalics without a functional hypothalamo-hypophyseal system.

Qu.Derom : Is anything known on the influence of treatment of corticoids in the mother on the morphology of the fetal endocrine pancreas?

An.Van Assche : Longstanding high-dose treatment with corticoids have been associated with intrauterine growth retardation. I have no personal experience on the effect of the fetal endocrine pancreas.

Qu.Moore : Is the rat a good morphological model for the human pancreatic cell types and their changes?

An.Van Assche : In previous publications we have clearly shown that there is a good analogy between the morphology of the endocrine pancreas between man and rat.

Qu.Teller : Farquhar stated that IDM tend to become more obese in later life than normal children. Also they are supposedly shorther than normal. Do these signs also show up in the IDM of rats in the second or third generation?
An.Van Assche : Indeed the offspring of diabetic rats are heavier 20 days after birth.

Qu.Moses : How do you explain the different results where under experimental conditions in fetal rat B-cells are not as responsive to hyperglycemia as the adult, and yet the fetal rat of a diabetic mother responds to hyperglycemia with increased insulin secretion?
An.Van Assche : It is so that the fetal B cell is not well responsive to a glucose stimulus. However, this is not the case when the glucose stimulus is of long duration.

ORAL FEEDING RECOMMENDATIONS IN FULL-TERM AND PREMATURE NEWBORNS

B. Salle, G. Putet, G. Meunier, J. Senterre

INTRODUCTION

Immediately after birth, the newborn breathes and expands its lungs, but the next requirement is food and water for maintaining normal plasma glucose and aminoacid levels, its body temperature and for growth. If food is not available just after birth, it is well known that the human newborn cannot maintain the deep body temperature, and supply energy and protein to the body cells. The full-term human baby is well endowed with fat, particularly brown fat and with glycogen in the liver. However, that will only suffice for a limited period of time, and the newborn must soon eat if he is to survive.

The aim of this presentation is to discuss oral feeding recommendations of full-term newborns and prematures, in the light of our knowledge of the digestion and absorption physiology, and the endocrine adaptation just after birth. Breast milk is considered as the basic nutrient for human babies, and the quality of formulas and other milks must be compared to it.

BREAST MILK

Breast feeding

It is now recommended for full-term infants and prematures, and few specific contra-indications exist (1). Ideally, breast milk should be practically the only source of nutrient for the newborn up to 4 or 5 months of life. Why should we breast feed babies?

		HUMAN MILK	NIDINA NESTLE	NURSIE GALLIA	S.M.A. WYETH	APTAMIL MILUPA	SIMILAC(MEAD) ENFANIL(ROSS)	ALPREM NESTLE
GLUCIDS	g							
Lactose	g	7	7.6	7.2	7.2	7.1	7.0	5.9
Glucose	g							2
LIPIDS	g	3.8	3.5	3.45	3.6	3.6	3.6	3.36
S.F.A./U.F.A.	%	46.2/53.8	47.5/52.5	48/52	44.1/55.9	51.4/48.5		
Linoleic ac.	g	0.31	0.42	0.44	0.46	0.36	—	—
Vegetable oils		0	+	+	+	+	+	+
M.C.T.	g	0	0	0	0	0	0	1.35
PROTEINS	g	1.2	1.7	1.7	1.5	1.7	1.5	2.02
Whey/Casein	%	60/40	53/47	50/50	60/40	55/45		70/30
MINERALS	mg	210	200	350	250	200	—	290
Ca	mg	33	47.2	57	40	65.6	55	54.6
P	mg	15	28.3	44	29	45	45	30.8
Ca/P		1.5-2	1.6	1.29	1.37	1.45	1.22	1.8
Na	mg	.15	16.2	26	16	20.2	5.5	16.8
Iron	mg	0.15	0.8	0.5	1.2	0.5	Traces	0.84
Copper	mcg	40	0	0	0	0	0.4	0.04
CALORIES		71	69	67	67	68.7	69	70

CONTENT FOR 100 ML

Table I

a. The milk of every mother (animal or human) is perfectly
adapted to the particular need of that species.
b. There are still differences (big and important) between
infant formulas and breast milk, despite newer "adapted" or
"humanized" infant formulas, which try to provide many of
the nutritional and physiological characteristics of breast
milk. The main differences between human milk and formulas
in carbohydrate,fat, protein and mineral contents are shown
in table I.

Protein
The total level of protein is generally less in human milk
than in formulas; moreover the quality differs considerably.
In adapted formulas whey/casein ratio is approximately
equal to 60/40. However human milk contains less aromatic
aminoacids, less methionine, more cystine and taurine which
has an influence on the blood levels of aminoacids (2 ,3).

In addition, human milk is rich in non protein nitrogen
in the form of nucleotides, urea and free aminoacids; the
total amount is approximately 25 %. This is a source of
non protein nitrogen which has been postulated as playing
a role in anabolism and growth (4 ,5). Recently, it has
been demonstrated that the breast milk of premature's mot-
hersis richer in nitrogen and mineral than in term's mothers so that nature is perfectly adapted to the newborn's
requirement (54) (Table II).

Fat and Cholesterol
Mean fat content of breast milk and formulas are the same.
However, fat content of breast milk varies considerably
from the beginning to the end of the suckling. In full-
term newborns, the absorption coefficient of human milk fat
is greater than 90 %. By contrast, it varies between 70
and 80 % even after one month of age with cow milk formulas
which have a fat content equal to or above 3 q/100Kcal
(6, 7, 8, 9). In all adapted formulas, vegetable oils replace
completely or partially butter fat, which improve fat absorption even in the first month of life. However, contrary to formulas, in human milk, the palmitic acid in esteri-

fied glycerol is mainly in position 2 (60 %) and this improve its digestibility (10).
Human milk has a linoleic acid content (7 to 12 % of total fat) which corresponds to between 3 to 6 % of the total caloric requirements. Some formulas contain more linoleic acid but no undesirable clinical effects are observed (11). Most of the cholesterol is removed with butter fat in formulas; this cholesterol might play a significant role in early feeding. Indeed, some authors think that exogenous cholesterol is necessary for formation of nerve tissue and for synthesis of bile salt (12,13). In addition, it could have an effect of dietary cholesterol on breast feeding, on the serum cholesterol level in early life and on the incidence of arterial sclerosis. On the other hand, long term effects of phytosterols provided by vegetable oils are not yet known.

TABLE II

NON PROTEIN NITROGEN IN MILKS
MG OF N/100 ML
(From FORSÜM et al)

	COW'S MILK	HUMAN MILK
Non Protein Nitrogen	28	50
Urea	13	25
Creatinine+Creatine	1,2	7,2
Uric Acid	0,8	0,5
α Amino Nitrogen	4,8	13
NH_4	0,6	0,2
Glucosamine	-	4,7

Calcium and Phosphorus
Calcium and phosphorous are lower in human milk than in formulas, but low calcium absorption is observed with formulas (7).

Iron
a. Iron content is low (2 to 3 mg/l) in human milk. However, 50 % of iron in human milk is absorbed and McMillon
reported that the iron in human milk is sufficient to meet
the iron requirement of breast fed full-term babies up to
6 or 7 months.
b. Iron supplementation in formulas does not prevent the
physiological fall of the red cell mass in the full-term as
in the premature, which is related to the shorter half-life
of foetal red cell rather than to an iron deficiency; full
term newborns fed from birth on formulas with an iron content of 12.5 mg/l do not have higher serum iron concentration nor a significant lower total iron binding capacity at
2 months of age than those receiving human milk or formulas
without iron (14). Lactoferrin is present in human milk in
much higher quantities than in formulas.

Immunologic Properties
It is evident that newborn infants can acquire important
elements of host resistance from breast milk while developping maturation of his own immunity system during the first
months of life.

Breast Milk Contains IGA which is resistant to proteolysis
and can confer passive mucosal protection of the gastrointestinal tract against penetration of intestinal organism
and antigens (15).

Lactoferrin has an inhibitory effect on Eschericha coli in
the intestin, its bacteriostatic effect is diminished as it
becomes saturated with iron (it is normally one third saturated by iron in human milk)(16).

<u>Lysozymes</u> which are bacteriostatic enzymes are much more abundant in human milk than in cow's milk. Bacterial lysis by IGA antibiotics does not occur unless lysozymes are present (17). Low concentration of specific complement fractions C3 and C4 in human milk are present but biologic importance is as yet unknown.

A possible intestinal immunity comes from the maintenance of a microflora with <u>lactobacillus</u>. The alimentary gut is sterile at birth and after 3 or 4 days more than 95% of the flora consists of the anaerobic lactobacillus with a minority of gram-bacteria and anaerobes (18). The mechanism by which a breast fed infant is able to maintain an acid stool with the lactobifidus as the predominant organism is poorly understood. Though most infant formulas provide a lactose content similar to that of human milk and a low buffering capacity, the predominant lactobifidus flora is not maintained.

Lastly, human milk spares the gastrointestinal tract from exposure to foreign food antigens and particularly to proteins, whilst a few micro-molecules can be absorbed and give local and general reactions. Some authors have brought some evidence of allergic manifestations later in childhood (eczema, asthma) in bottle fed infants, because of early exposure to cow's milk antigens and proteins (19,20,21).

FULL-TERM NEWBORN

Breast feeding is strongly recommended for the full-term infant. It is very important in our countries for all physicians, obstetricians and paediatricians to become more knowledgeable about infant nutrition and value the technique of breast feeding.

As for the opportunity for successful lactation and breast feeding, it is necessary and important.

1. To avoid separation of the mother from her infant during the first 24 hours. Furthermore, to decrease the amount of sedation and anesthesia to the mother during labor. A large amount of anesthesia can impair suckling of the baby.

2. To allow suckling just after birth: early stimulation of the mammary glands after birth appears to encourage
successful lactation. The sensitive period for this is in
the minutes and hours following delivery. Suckling is a
very important stimulus to the release of prolactin and
oxytocin.

The first food is colostrum and this nutrient is different
in composition from later milk in many ways : composition of
protein, fat and carbohydrate in colostrum is seen in Table
III. for milk comparison in four species, and the greatest
difference is in the concentration of protein. Colostrum
has much more protein than mature milk and some of this is
in the form of immunoglobulins and lactoferrin, protecting
against infection (Ogra et al (22,23). The concentration
of immunoglobulins in colostrum ranged from 1000 to 1500mg
per 100 ml for IGA, 100 to 250 mg per 100 ml for IGM and
40 to 50 mg for IGG and decreased within 5 days. Parallely
the levels of immunoglobulins in the serum did not manifest
any significant change particularly during the first days
of life; so far colostrum contributes largely to intestinal
immunity. 50 to 10 M of IGA immunoglobulins are absorbed,
from birth to 24 hours, but after 36 hours, neonates fail
to absorb any IGA or other immunoglobulins (22,23). Furthermore, a large number of lymphocytes and macrophages were
observed in the colostrum specially T lymphocytes. These
data suggest that colostrum may be crucial in providing the
infant with a highly concentrated bolus of immunoglobulins
early (22). Sodium and vitamin A are higher in colostrum
than in human milk; immaturity of the kidney in the neonatal period is known to impair the retention of sodium.

TABLE III

From WIDDOWSON E.M., in Gastrointestinal Development
and Neonatal Nutrition, 1977. Composition of Colostrum (C)
and Mature Milk (M) of four Species, G per 100 ml.

		WOMAN	COW	MARE	SOW
Protein	C	2.7	14.2	7.2	17.8
	M	1.2	3.3	2.2	5.8
Fat	C	2.9	3.6	2.5	4.4
	M	3.8	3.7	1.6	8.5
Lactose	C	5.3	3.1	4.7	3.5
	M	7.0	4.8	6.0	4.8

Breast feeding must start early after birth to avoid the
metabolic disorder of starvation. In addition, early feeding develops the secretion of all gastrointestinal hormones
as shown by Lucas et al. in Oxford, and the secretion from
exocrine intestinal cells and pancreas particularly (55).
It must be on demand rather on a rigid 3 to 4 hours schedule.
If breast milk is not available, formulas are preferable to
cow's milk in the case of full-term babies (mother's refu-
sal or contra-indication).

PREMATURES

Feeding regimen for premature infants is to obtain first a
prompt postnatal weight gain and secondly to maintain a
certain weight increase whilst keeping a biologic equili-
brium(24). The principal needs for premature infants are
as follows :

1. Energy requirements of premature infants is lower than
that of full-term newborns during the first week of life
but it reaches and ever exceeds that of the full-term newborn in the second week of life. In addition, if we take
into account the growth rate, caloric requirements must
increase daily to reach an average of 120-130Kcal/kg/day
(25).

2. Water requirements depend on many factors (gestational
age of the babies more than the weight (Hull), osmotic load
of the milk or formulas, capacity of the immature kidney
for concentrating urine, exposition of the premature to
heat, phototherapy, cold stress, infection, diarrhea,etc...).
A daily intake of 170-180 ml/kg of water is generally accepted but not always sufficient in certain conditions particularly in very low birth weight infants (26).

3. The optimal requirement of protein intake for the premature infants has not been precisely defined; it is as recommended by Räihä et al. from 2 to 2.5 g/kg/day (2,3), by Senterre from 3 to 3.5 g/kg/day (27). During the first 8 weeks
of life according to maturity of birth (gestational age),

aminoacid synthesis and protein degradation capacities may
not yet be fully developped. A compromise must be made
between estimated requirement and actual intake in order to
avoid metabolic imbalances which may have deleterious effects on brain development and later on, on intellectual
performance. So it is very important in the premature infant to take into account not only the quantity, but also
the quality of protein to provide enough of the aminoacids
which are not being synthetised by the infant, and less of
the aminoacids which are not well metabolised by the liver.

Räihä et al. (2,3,28) in many reports fed premature infants with five formulas including pooled breast milk. The
breast milk supplied approximately 1.7 g/kg/day, two formulas 2.25 g/kg/day and 2 other 4.5 g/kg/day of protein; formulas differ with whey/casein ratio. All infants grew
equally well when fed 117Kcal/kg/day. Statistically the
breast fed group gained at a slightly lower rate; but
significant difference in plasma aminoacids ammonaemia,
and urea levels were noted in the babies. The lower ammonaemia, tyrosine and phenylalanine levels were found in
infants fed with human milk or in those fed low protein
formula with whey/casein ratio of 60/40.

4. Fat : the ability of premature infants to absorb fat,
particularly saturated fatty acids is relatively poor (7,
29). This limitation is associated with decreased bile
salt synthesis and low pancreatic lipase secretion. It is
recommended to feed premature babies either with low fat
formulas or with vegetable oils, rich in unsaturated fatty
acids (figure 1.). Medium chain triglycerides as part of
fat (40 %) has been shown to improve fat absorption in premature infants (30,31); they increase weight by increasing
caloric absorption, they enhance calcium absorption and
nitrogen retention (Table IV). The fat intake is about
6 g/kg/day that means 50 % of the caloric requirement of
the preterm infant.

5. <u>Carbohydrate</u>. Lactose is the natural sugar of human milk. It has been generally added to cow's milk formula to increase the carbohydrate content up to that of human milk. However prematures have sometimes a delay in the maturation of intestinal lactase and this may be of physiological consequence.

Figure 1. : Net absorption of fatty acids in preterm infants fed human milk (HM), various adapted formulas (AF1, AF2, AF3) or half-skimmed cow's milk (J.Senterre, 1976) (27).

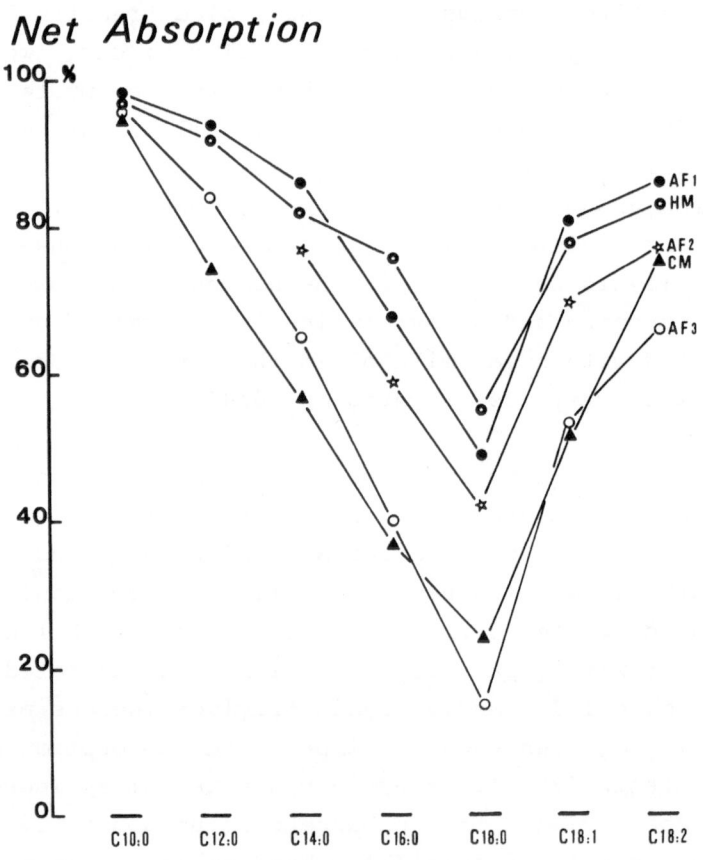

In a recent study, Fosbrooke (56) has found that the addition of sugar . other than lactose to the milk resulted in
a lower incidence of diarrhea and metabolic acidosis.

6. <u>Calcium and Phosphorus</u> are high in preterm infants.
Calcium content of human milk is low but it is well absorbed about 50 %. However human milk appears to be deficient
in the phosphorus required to fulfill both cellular growth
and bone mineralization. In these conditions, phosphorus
is preferentially used for building up protoplasm rather
than for skeletal calcification and even though the amount
of calcium absorbed is high, one part of it is excreted in
the urine. Therefore, to avoid that hypercalciuria, it is
justifiable to supplement banked human milk with phosphate
(Senterre et al. (27,31)).

TABLE IV

NITROGEN, FAT, CALCIUM AND PHOSPHORUS BALANCES IN 12 PRETERM
INFANTS FED A ESPECIALLY ADAPTED FORMULA WITH WHEY-CASEIN RATIO
70/30 AND MEDIUM CHAIN TRIGLYCERIDES (40 % OF FAT).
(J. Senterre, unpublished data)

	NITROGEN mg/kg/d	FAT g/kg/d	CALCIUM mg/kg/d	PHOSPHORUS mg/kg/d
INTAKE	594 ± 30	6.2 ± 0.1	89 ± 6	63 ± 5
FECES	45 ± 18	0.7 ± 0.3	37 ± 8	4 ± 2
URINE	158 ± 33	-	3 ± 2	19 ± 2
RETAINED	391 ± 39	5.5 ± 0.4	49 ± 10	40 ± 5
ABSORPTION (%)	92 ± 3	89 ± 6	58 ± 9	94 ± 3
RETENTION (%)	66 ± 5		55 ± 10	63 ± 5

In formulas calcium content is higher but the amount of calcium which is absorbed is not much higher than that of human
milk. In adapted formulas, if the calcium content is too
close to that of human milk, calcium balances may be nega-
tive, but if calcium content is greatly increased, fat
absorption may be impeded, resulting in a substantial fecal
loss of calories. However, to avoid an overload in phosphorus, the Ca:P ratio in adapted formulas should be around
1.8 (27,32) (Table V).

Table V : Calcium and phosphorus balance data in preterm
          infants fed human milk (HM), various adapted for-
          mulas (AF1,AF2,AF3) or half-skimmed cow's milk
          (J. Senterre, 1978)(31).

TABLE V

	HM	AF1	AF2	AF3	CM
Calcium					
Intake (mg/kg/day)	$55^{\pm}9$	$71^{\pm}9$	$90^{\pm}5$	$129^{\pm}5$	$199^{\pm}14$
Feces (mg/kg/day)	$28^{\pm}9$	$62^{\pm}12$	$60^{\pm}13$	$101^{\pm}11$	$161^{\pm}17$
Urine (mg/kg/day)	$10^{\pm}5$	$3^{\pm}1$	$2^{\pm}1$	$3^{\pm}1$	$2^{\pm}1$
Retained (mg/kg/day)	$17^{\pm}9$	$6^{\pm}15$	$28^{\pm}12$	$25^{\pm}11$	$36^{\pm}13$
Absorption (%)	$49^{\pm}11$	$13^{\pm}21$	$33^{\pm}14$	$22^{\pm}8$	$19^{\pm}6$
Retention (%)	$31^{\pm}14$	$9^{\pm}22$	$31^{\pm}14$	$19^{\pm}8$	$18^{\pm}7$
Phosphorous					
Intake (mg/kg/day)	$29^{\pm}4$	$65^{\pm}7$	$59^{\pm}4$	$82^{\pm}4$	$163^{\pm}8$
Feces (mg/kg/day)	$3^{\pm}1$	$7^{\pm}4$	$5^{\pm}2$	$5^{\pm}2$	$57^{\pm}9$
Urine (mg/kg/day)	$1^{\pm}1$	$37^{\pm}10$	$23^{\pm}9$	$52^{\pm}8$	$68^{\pm}10$
Retained (mg/kg/day)	$25^{\pm}5$	$21^{\pm}7$	$31^{\pm}9$	$25^{\pm}9$	$38^{\pm}11$
Absorption (%)	$89^{\pm}5$	$89^{\pm}4$	$92^{\pm}2$	$94^{\pm}2$	$65^{\pm}9$
Retention (%)	$87^{\pm}8$	$32^{\pm}10$	$53^{\pm}15$	$31^{\pm}10$	$23^{\pm}7$

7. Vitamins : two vitamins are mainly considered :

- Vitamin D : it is generally noted that very low birth weight infants cannot metabolize vitamin D into its metabolites and some advocate giving 25OHD3 or 1.25 (OH)$_2$ D3. But our results demonstrated that prematures can hydroxylate in the liver and the kidney vitamin D into its active metabolites (33,34). According to the general vitamin D status of mothers in our countries, the recommended amount of vitamin D is 1.000 I.U. during Summer and 2.000 I.U. during Winter.

- Vitamin E : vitamin E requirement merits special consideration. Absorption of vitamin E is poor in the prematures fed with formulas; the plasma level of tocopherols is low in comparison to the babies fed with human milk. The requirement for vitamin E increases as the level of polyunsaturated fat in the diet increases. When the iron levels in formulas are high, the requirement for vitamin E by low premature infants also increases. It is recommended to give 5 I.U. or mg to low birth weight infants/day and not to exceed 25 I.U. or mg/day during the first 3 months of life. After 3 months, vitamin E supplementation should not be necessary (35).

We have to consider now the practical point of view of feeding   in prematures :

a. Prematures weighing more than 1.200 g : there is no major problem in feeding a premature by gavage through a nasogastric tube or by suckling as soon as possible. But if the baby weighs less than 2 kilos, it is also necessary and important to perfuse the baby with a mixture of glucose, aminoacids and electrolytes as recommended by many authors until he receives an amount of 80 Kcal/kg/day of milk (36, 37,38). With such a method, he gains his birth weight more rapidly without disturbances of calcium, sodium and glucose levels. In sick prematures (under assisted ventilation, or with sepsis,etc.) the same principles can be applied, but oral nutrition is generally delayed two or three days after birth.

It is desireable to use mainly human milk from the

mother or from a pooled bank of milk. If more protein is
required, the breast milk should be supplemented preferably
with whey protein hydrolysates which have no antigenic properties and a high biological value. But the time and the
amount of this supplementation should be adjusted according
to the schedule of biochemical investigation. As Räihä
pointed out (2,3), although a slight growth advantage in
terms of weight gain is achieved with early increase of protein administration, it is not certain such an increased
growth rate is desirable. We are not yet sure if we have to
maintain the same intrauterine rate of weight gain during
the immediate extrauterine life.

In addition to the problem of protein, it has been
clearly demonstrated that human milk must be supplemented
in phosphorous and sodium in the form of disodic phosphate
to avoid hypercalciuria and sodium depletion (32).

If human milk is not available, it is necessary to
use a formula. Recent work has shown that specially adapted formulas for preterm infants with high whey / casein
ratio medium chain triglycerides, calcium/phosphorous ratio
about 2, relatively low lactose and glucose or malto-dextrin
content, have some advantages in comparison to usual infant
formulas.

b. Prematures weighing less than 1.200 g : the small
preterm infant of 1.200 g or less is in a very special situation where his enormous energetic needs for maintenance
and growth are challenged by the fact that he is not fully
adapted for an extrauterine life.

The main problems are the following :
- temperature regulation
- water balance adaptation which is impaired by higher basal
insensible water losses; furthermore this is increased by
phototherapy or nursing under a radiant heat warmer.
- electrolytes imbalance which are mainly due to renal immaturity with increased bicarbonates and sodium losses due to
a low glomerular filtration rate and an inadequate renal
concentration ability.

- glucose homeostasis with greater risk of hyperglycemia
and glycosuria with hyperosmolarity. These are the results
of a decreased glucose tolerance by impairment of insulin
secretion (39,40,41).

These basic maintenance problems are most often complicated by the special and frequent pathology of these
tiny newborns, such as :
- respiratory distress syndrome with its ventilation management problems (50 %)
- cardio-vascular problems such as patent ductus arteriosus
resurgence leading to therapeutic fluid restrictions.
- infections
- gastro-intestinal intolerance and/or necrotizing enterocolitis.

Thus, special nutritional requirements are needed to
maintain growth. The frequent inability of naso gastric
alimentation during the first days or weeks of life have led
to a search for alternate routes of feeding : mainly parenteral (either as total parenteral nutrition or as parenteral
supplementation) (42,43,44,45,46,47) or nasojejunal alimentation (48,49,50).

Total parenteral alimentation has the main advantage of
a higher caloric intake quickly obtained; it can be done by
progressive increase of carbohydrate concentration in fluids
given through a central line and by the use of fat emulsion
through a peripheral vein. It usually results in a quick
gain of weight. Metabolic disorders (51,53), infections
and catheter complications are the most common disadvantages
of this method. Furthermore, recent date from Lucas (55)
showing a delayed secretion of gastrointestinal hormones
with absence of oral feeding seems to be of importance.
However, this delayed secretion does not seem to matter in
parenteral supplementation. Nasojejunal continuous alimentation has the main advantage of avoiding gastric emptying
problems; few, if any metabolic disorders occur with this
method but complications such as intestinal perforation (52)
with polyvinyl catheter, impaired nutriments absorption and
necrotizing enterocolitis have been described.

In our department (Lyon), we have combined I.V. alimentation through a central line and nasojejunal feeding.

34 very premature newborns (birth weight less than 1.200 g) were managed with this method over 1 year; 19 of them had to be ventilated and 22 of 34 survived; clinical data are given in Table VI.

TABLE VI

CLINICAL DATA

INFANTS	NUMBER	WEIGHT (g) (mean±SD)	G.A. (w) (mean±SD)	R.D.S. WITH ASSISTED VENTILATION	SURVIVALS
BIRTHWEIGHT < 1000	16	832.5 ± 52	28.2 ± 1.2	10	8 (50%)
BIRTHWEIGHT 1001-1200	18	1136.4 ± 54.7	30.5 ± 2.2	9	14 (77%)
	34			19	22 (65%)

At admission of the newborn, a central line is inserted (as we think that a peripheral perfusion has the disadvantage of being replaced very often, implying disturbances of a tiny newborn in an instable condition). At the same time or during the first 24 hours a nasojejunal tube is inserted under T.V. control; this feeding tube is replaced every 4 days as it may become rigid and is never manipulated.

During the first day only I.V. alimentation is performed at a rate of 100 ml/kg/day. It is a mixture of cristalline aminoacids, carbohydrate and minerals as noted in Table VII.

347

TABLE VII

INTRAVENOUS INTAKE/KG/DAY

Nitrogen	170	mg
Na	5	mEq
Fructose	2	g
Glucose	5	g
Ca	35	mg
K	0,35	mEq

Nasojejunal feeding, using only human milk, is usually started during the second day, infused at a continuous rate using a constant infusion pump. Total fluid intake reaches progressively 170-180 ml/kg/day by the fifth day as seen on Figure 2; during this period I.V. intake is stable as oral feeding accounts for most of the daily rise. I.V. alimentation is usually stopped on the tenth day of life, at which time newborns receive 120-140 ml/kg/day of human milk and are in a growing state.

Nasojejunal alimentation is usually stopped by the twelveth day and conventional nasogastric feeding started.

With this feeding method, the usual maintenance caloric

intake (70-80 Kcal/kg/day) is achieved by the end of the
first week of life (Fig.3) when the newborns start gaining.
At that time about half of the fluid volume and half of the
energy daily intake is given by milk as it is seen in figure 3. The birth weight is usually regained within 15
days.

The results are seen in Table VIII out ^{of} 34 newborns, 11 died
before 48 hours (eight of them weighing less than 1.000 g)
and before nasojejunal alimentation. Infectious complications are listed in Figure 4; it can be seen that most of
the venous catheters were sterile and that there was no
sepsis nor necrotizing enterocolitis in our series. We
have to precise that non prophylactic antibiotic are used.
Metabolic complications are usual disorders during the
first days of life in the very low birth weight babies and
they happened in our series as shown in Table IX.

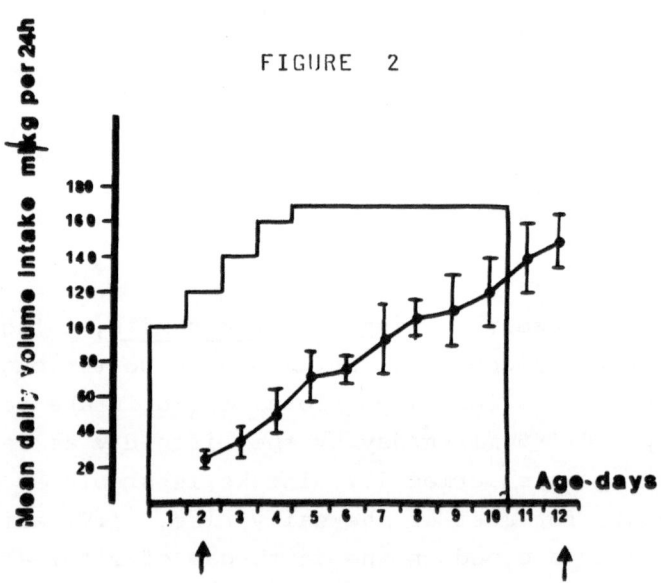

FIGURE   2

Total volume intake and mean milk intake
in 22 infants

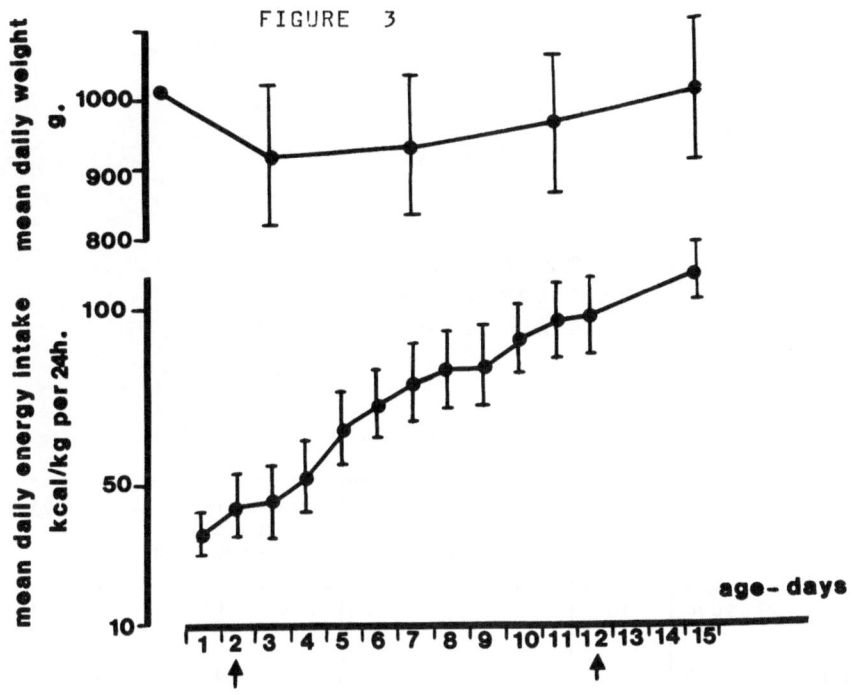

FIGURE 3

Out of 22 survivals we noted particularly glucose and
sodium disturbances and the most frequent was hypoglycemia;
carbohydrate intake shown in Figure 5 was very low during
the first 24 hours, during which most of these hypoglycemias
occur ; but these babies usually received transfusions
which decreased glucose intake as the total fluid intake
was maintained at 100 ml/kg/day including blood, and some

babies were very ill on admission. It has to be noted that
we never observed hyperglycemia.

TABLE VIII

RATE OF SURVIVALS OF INFANTS   1200 G
DURING STUDY PERIOD
(October 1978-October 1979)

INFANTS	NUMBER	DIED BEFORE 48 H OF LIFE (Before alimentation)	DIED DURING NASO-JEJUNAL ALIMENTATION	SURVIVALS
BIRTHWEIGHT < 1000	16	8	0	8 (50%)
BIRTHWEIGHT 1001-1200	18	3	1	14 (78%)
TOTAL	34			22 (65%)

Sodium disturbances included hypernatremia which again
happened during the third or fourth day of life when the
weight was at its lowest.

We never observed acid-base balance disorders, but all
our babies received bicarbonate supplementation.

INFECTIOUS COMPLICATIONS

- Necrotizing enterocolitis   0

- Sepsis   0

- Cultures
  - . Venous catheter (n23)
    Sterile   21
    Staphylococcus Epidermidis   2

  - . Jejunal tube (n50)
    Streptococcus faecalis   3
    Staphylococcus Epidermidis   7
    Enterobacter Faecalis   1
    Bacillus   2

FIGURE 4

TABLE IX

METABOLIC DISTURBANCES DURING
THE STUDY PERIOD (22 INFANTS)

GLUCOSE (mmol/l)	>	7.5	0
	<	1.5	9
SODIUM (mmol/l)	>	150	5
	<	130	2
POTASSIUM (mmol/l)	>	8	0
	<	3	2
CALCIUM (mmol/l)	>	2.5	0
	<	1.7	4
BILIRUBIN (umol/l)	>	200	5
PROTEIN (g/l)	<	40	0
UREA (mmol/l)	>	9	1

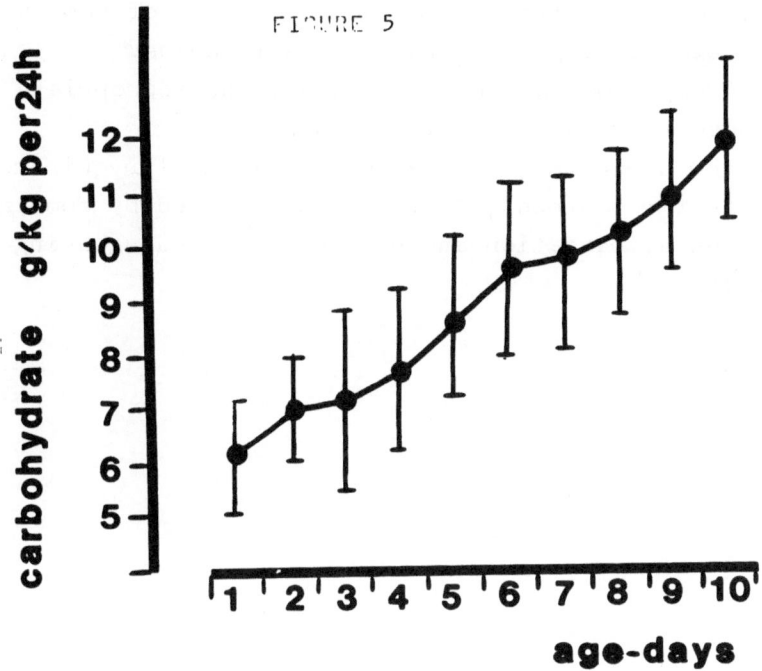

FIGURE 5

CONCLUSION AND SUMMARY

Breast feeding is above all recommended for full-term or premature newborn; except if there are specific contra-indication or/and when breast feeding is unsuccessful.

In full-term newborn, adapted formula can replace without major disturbances, human milk except for immuno-logic properties.

In premature infants, fresh breast milk is the best nutrient but it is necessary to increase the phosphate and the protein content as indicated above. The apparent nor-mal growth of prematures fed breast milk which supplies 1.7 to 2 g/of protein per kg of body weight may be caused by its distribution of aminoacids (2). Formulas may be used but they should include vegetable oils and medium chain triglycerides which are well absorbed and enhance the absorption and retention of nitrogen and certain mine-rals. The supplementation of the diet with vitamin E is particularly important because of poor absorption by the premature of this vitamin and the occurrence of hemolytic anaemia related to polyunsaturated fatty acids.

Vitamin D supplementation is necessary and appropriate at higher dose than in the full-term newborn.

In very low birth weight infants, nutritional requirements during the first weeks of life can be covered by combined intravenous alimentation and nasojejunal feeding using only human milk.

REFERENCES
1. Nutrition Committee of the Canadian Paediatric Society
    and the Committee on Nutrition of the American Acade-
    my of Pediatrics. Breast Feeding:Pediatrics 62:591-
    601,1978
2. Räihä, NCR, K Heinonen, DK Rassin, GE Gaull, Milk pro-
    tein quantity and quality in low-birth weight infants.
    I. Metabolic responses and effects on growth. Pedia-
    trics 57:659, 1976
3. Gaull, GE, DK Rassin, NCR Räihä, K Heinonen, Milk pro-
    tein quantity and quality in low birth weight infants:
    III. Effects on sulfur amino-acids in plasma and urine.
    J.Pediatr 90:348, 1977
4. ESPOAN Committee on Nutrition, Guidelines on Infant
    Nutrition,I. Recommendations for the composition of
    an adapted formula. Acta Paediatr   262:5-20,
    1977
5. Hambraeus, L, E Forsum, B Lonnerdal, Nutritional as-
    pects on breast milk versus cow's milk formula sympo-
    sium on nutrition in early life, Lund, 1975
6. Hanna, FM, DA Navarette, FA Hsu, Calcium-fatty acid
    absorption in term infants fed human milk and prepared
    formulas simulating human milk. Pediatrics 45:216,
    1970
7. Shaw, JCL, Evidence for defective skeletal mineraliza-
    tion in low-birth weight infants : the absorption of
    calcium and fat. Pediatrics 57:16, 1976
8. Widdowson, EM, Absorption and excretion of fat, nitro-
    gen and minerals from "filled" milks by babies one
    week old. Lancet II:1099, 1965
9. Widdowson, EM, Preparations used for the artificial
    feeding of infants. J R Coll Physicians Lond 3:285,
    1969
10. Rey, J, C Ricour, La spécificité moléculaire de l'ab-
    sorption des graisses. Biol Gastroenterol (Paris)
    5:187, 1972
11. Ballabriga, A, A Martinez, A Gallart-Catala, Composi-
    tion of subcutaneous fat depot in prematures in rela-
    tionship with fat intake. Helv Paediatr Acta 27:91,
    1972
12. Petrich, E, W Plenert, The concentration of essential
    fatty acids in human milk. Acta Paediatr Scand
    64:153, 1975
13. Ziegler, E, Modern trends in composition of infant
    formulas. Symposium on Nutrition in Early Life.
    Lund, 1975
14. Pearson, HA, Life-span of the fetal red blood cell.
    J Pediatr 70:166, 1967
15. Hanson, LA, J Winberg, Breast milk and defense against
    infection in the newborn. Arch Dis Child 47:845, 1972
16. Saarinen, UM, MA Siimes, PR Dallman, Iron absorption
    in infants : High bioavailability of breast milk iron
    as indicated by the extrinsic tag method of iron ab-
    sorption and by the concentration of serum ferritin.
    J Pediatr 91:36, 1977
17. Goldman, AS, CW Smith, Host resistance factors in hu-
    man milk. J Pediatr 82:1082, 1973

18. Mata, LJ, ML Mejicanos, F Jimenez, Studies on the in-
    digenous gastrointestinal flora of Guatemalan children.
    Amer J Clin Nutr 25:1380, 1972
19. Gerrard, JW, JWA MacKenzie, N Goluboff et al, Cow's
    milk allergy : Prevalence and manifestations in an un-
    selected series of newborns. Acta Paediatr Scand
    suppl 234, 1973
20. Matthew, DJ, B Taylor, AP Norman et al, Prevention of
    eczema. Lancet I:321, 1977
21. Easthman, EJ, T Lichauco, MI Grady, WA Walker, Anti-
    genicity of infant formulas. Role of immature intes-
    tine on protein permeability. J Pediatr 93:561, 1978
22. Ogra, SS, PL Ogra, Immunologic aspects of human colos-
    trum and milk. I. Distribution characteristics and
    concentrations of immunoglobulins at different times
    after onset of lactation. J Pediatr 92:458, 1978
23. Ogra, SS, PL Ogra, Immunologic aspects of human colos-
    trum and milk. II. Characteristics of lymphocytes
    reactivity and distribution of E. Rosette forming
    cells at different times after the onset of lactation.
    J Pediatr 92:550, 1978
24. Committee on Nutrition : Nutritional Needs of Low-birth
    Weight Infants, American Academy of Pediatrics,
    Pediatrics 60:519-530, 1977
25. Sinclair, JC, JM Driscoll, Jr, WC Heird, RW Winters,
    Supportive management of the sick neonate : parente-
    ral calories, water, and electrolytes. Pediatr Clin
    North Am 17:863, 1970
26. Gordon, HH, SZ Levine, The metabolic basis for the in-
    dividualized feeding of infants, premature and full-
    term. J Pediatr 25:464, 1944
27. Senterre, J, L'alimentation optimale du prématuré.
    Ed. Vaillant Carmanne Liège, 1976
28. Räihä, NCR, Biochemical basis for nutritional manage-
    ment of preterm infants. Pediatrics 53:147, 1974
29. Gordon, HH, H McNamara, Fat excretion of premature in-
    fants: I. Effect on fecal fat of decreasing fat in-
    take. Am J Dis Child 62:328, 1941
30. Roy, CC, M Ste Marie, L Chartrand, A Weber, H Bard,
    B Dorey, Correction of the malabsorption of the pre-
    term infant with a medium-chain triglycerides formu-
    la. J Pediatr 86:446, 1975
31. Senterre, J, Calcium and phosphorous retention in pre-
    term infants. Intensive care in the Newborn II,
    L. Stern, B.Friis-Hansen and W.Oh editors, publ.
    Masson New York 205, 1978
32. Senterre, J, Unpublished data
33. Glorieux, F, BL Salle, E Delvin, L David, Serum 25-
    hydroxyvitamin D (25-OHD) levels following vitamin D
    administration during the first week of life in pre-
    mature infants. Communications à l'American Society
    for Pediatric Research, Atlanta, 1-5 May 1979 (U.S.A.),
    Abstract in Pediatr Res 13:475, 1979
34. Glorieux, F, BL Salle, L David, E Delvin, 1.25 (OH)$_2$
    D3 in preterm infants treated or not treated by vita-
    min D. Soumis pour publication New Engl J Med 1980

35. Dallman, PR, Iron Vitamin E and Folate in the preterm infant. J Pediatr 85:742, 1974
36. Bryan, MH, P Wei, JR Hamilton, GW Chance, PR Swyer, Supplemental intravenous alimentation in low birth weight infants. J Pediatr 92:940, 1973
37. Pildes , RS, RS Ramamurthy, GV Cordero, PWK Wong, Intravenous supplementation of L. aminoacids and dextrose in low birth weight infants. J Pediatr 82:945, 1973
38. Salle, BL, M Vercherat, Alimentation intraveineuse supplémentaire chez le prématuré de moins de 1500 g. Arch franç Pediat 32:27-37, 1975
39. Salle, BL, GW Chance, A Ruiton-Ugliengo, Glucose assimilation in small prematures. In J.Stetson and P.R. Swyer, Neonatal Intensive Care 165 (Warren H. Green, St Louis Mo, 1974)
40. Salle, BL, A Ruiton-Ugliengo, Glucose disappearance rate, insulin response and growth hormone response in the small for gestational age and premature infant of very low birth weight. Biol Neonate 29:1-17, 1976
41. Dweck, HS, G Cassady, Glucose intolerance in infants of very low birth weight. Incidence of hyperglycemia in infants of birth weights 1100 grams or less. Pediatrics 53:189, 1974
42. Driscoll, JM, Jr, WC Heird, JN Schullinger, RD Gongaware, RW Winters, Total intravenous alimentation in low birth weight infants. A preliminary report. J Pediatr 81:145, 1972
43. Hall, RT, PG Rhodes, Total parenteral alimentation via undwelling umbilical catheters in the newborn period. Arch Dis Child 51:929, 1976
44. Yu, VY, B James, P Hendry, RA MacMahon, Total parenteral nutrition in very low birth weight infants, a controlled trial. Arch Dis Child 54:653, 1979
45. Brans, YW, JE Sumners, HS Dweck, G Cassady, Feeding the low birth weight infant : orally or parenterally I. Preliminary results of a comparative study. Pediatrics 54:15, 1974
46. Brans, YW, JE Sumners, HS Dweck, PE Dailey, G Cassady, Feeding the low birth weight infant, orally or parenterally. II. Corrected bromide spare in parenteral supplemented infants. Pediatrics 58:809, 1976
47. Peden, VM, JT Karpel, Total parenteral nutrition in premature infants. J Pediatr 81:137, 1972
48. Cheek, JA, Jr, GF Staub, Nasojejunal alimentation for premature and full-term newborn infants. J Pediatr 82:955, 1973
49. Van Callie, M, GK Powell, Nasoduodenal versus nasogastric feeding in the very low birth weight infants. Pediatrics 56:1065, 1975
50. Rhea, JW, O Ghazzani, N Weidman, Nasojejunal feeding; an improve device and intubation technique. J Pediatr 82:951, 1973
51. Heird, WC, JF Nicrolson, JM Driscoll, Jr, JN Schullinger, RW Winters, Hyperamnonemia resulting from intravenous alimentation using a mixture of synthetic L aminoacids. A preliminary report. J Pediatr 81;162, 1972

52. Boros, ST, JW Renolds, Duodenal perforation. A compli-
    cation of nasojejunal feeding. J Pediatr 85:107, 1973
53. Rigo, J, J Senterre, J Oger, Improvement of neonatal
    growth by parenteral nutrition in low birth weight
    infant. Acta Paediatr Belg , 1979
54. Atkinson, SA, MH Bryand, IC Radde, et al, Effect of
    premature birth on total nitrogen and mineral concen-
    tration in human milk. Read before the Western
    Hemisphere Nutrition Congress V, Quebec City, August
    1977
55. Lucas, A, TE Adrian, SR Bloom, A Aynsley-Green, Gut
    hormones in the neonate. Pediatr Res 14:177 (abstract)
    1980
56. Fosbrooke, AS, BA Wharton, "Added lactose" and "added
    sucrose" cow's milk formulas in nutrition of low
    birth weight babies. Arch Dis Childh 50:409, 1975

DISCUSSION ABOUT THE FOURTH SESSION

Qu.Eggermont : How do you manage hyperglycemia induced by
feeding very tiny babies?

An.Salle : In very tiny babies, we look at plasma levels of
glucose every 6 hours. First of all, we never use insulin
in the Department, as recommended by some authors and particularly by Senterre (Personal communication).
Secondly, as you can see in Table IX, we never have observed hyperglycemia; the reason is that we never exceed an
amount of glucose of 7 mg/kg/min by infusion.
We always use amino-acids from birth on and as shown by
Grasso et al. (Diabetes, 1970, 19,837), Reitano et al.
(Diabetes, 1978,27,334), amino-acids and theophylline probably enhance the secretion of insulin by the pancreas.

Qu.Teller : Could you comment on the role of vitamin E in
the feeding of premature infants?

An.Salle : The role of vitamin E has been defined by Peter
Dallmann in Journal of Pediatrics (35). The evidence of
vitamin E deficiency syndrome in preterm infants is actually evident and well documented (Oski and Barness, Journal
of Pediatrics, 1967,70,211; Ritchie JH,Fish MB,Mc Master U
and Grossman, New England Journal of Medicine, 1968,269,
1185). Vitamin E plays a role in the synthesis of heme
facilitating dehydrase reactions.
The requirement of vitamin E depends on the degree of saturation of the fatty acids and of the iron in the diet.
Diets particularly rich in polyunsaturated fatty acids produce a change in the composition of fatty acid in cellular
and intra-cellular membranes, which become more susceptible
to damage as a result of lipid deoxydation. This effect
needs the antioxydant effect of vitamin E and this is seen
particularly in prematures fed with human formulas.
Melhorn et al. (Blood, 1971,37,438) pointed out that infants
who received only iron (8 mg/kg) were not significantly
more anemic than infants receiving vitamin E and had higher
reticulocyte counts suggesting the possibility of hemolysis.
Iron probably catalyses the oxydative breakdown of red cell

lipids which is less pronounced if vitamin E exerts its
antioxydant effects. There is evidence that concurrent
administration of iron and vitamin E impairs absorption of
vitamin E (α tocopherol).

The management of vitamin E administration in the first 12
weeks of life is complicated by the poor assimilation of α
tocopherol acetate which is the available form of vitamin
A dose of 25 IU of α tocopherol acetate per day does not
raise considerably the vitamin E concentration in the plasma to the normal range (Melhorn, Gros and Childers, G,
Journal of Pediatrics, 1971,79,581) but the absorption of
vitamin E can be facilitated by water soluble α tocopherol
polyethylene glycol succinate; a dose of 10 to 15 IU is
likely to be adequate during the first 3 months of life.

Qu.Shelley : Why did you include fructose in the parenteral
feeding of very small babies?
An.Salle : We did not supplementate fructose in the parenteral feeding, but we perfuse the baby with a formula containing fructose namely (vamin-Vitrum).

Com.R. De Meyer : It seems actually that fructose has a
depressing effect on liver aldolase. Therefore one should
be very cautious using this sugar in newborn babies.

Qu.Eggermont : The statement of Professor De Meyer on a
depressed liver aldolase activity in premature infants also
worries us. Did you observe a more pronounced fall of plasma phosphate when giving fructose?
An.Salle : We did not observe any fall of plasma phosphate,
nor acidosis (Salle et al.,Archives Françaises de Pédiatrie
(38), Rigo and Senterre, Acta Paediatrica Belgica (53)).
Nevertheless we supplementate the perfusion with phosphorous
diphosphate and we never exceeded an amount of 4 g/kg/day
of fructose intravenously.

Qu.Van Assche : The very small infant probably needs insulin?

An.Senterre : In the infants below 1200 g, hyperglycemia
may be observed even with low amounts of infused glucose,
especially during the first days of life because of a lack
of insulin secretion. In these circumstances, if the rate
of infused glucose must be decreased to such an extent that
the caloric supply becomes inadequate, we increase the glucose tolerance by a continuous infusion of insulin at a
rate of 0.1-0.3 unit/kg/hour from a solution of 6 units of
insulin per ml in 2 % of human albumin. Of course, this is
not done in case of acute hyperglycemia related to a peri-
pherical resistance to insulin as it can be seen in severe
infection.

Com.Verellen : We can confirm that babies fed by their
own mother's breast milk are receiving 3 g/kg/d and do
retain 300 mg of N per kg per day.

An.Senterre : I agree with you that nitrogen balance can
reach 300 mg/kg/day in preterm infant fed human milk if it
is his own mother's breast milk. Indeed as you know it has
been demonstrated that milk from mothers of a premature baby
has a protein content about 25 % higher than that of banked
human milk. From the results of our metabolic balance
studies, nitrogen retention is about 260 mg/kg/d in premature infants fed 190 ml/kg/d of banked human milk but nitrogen balance can be improved by increasing nitrogen intake
(J.Senterre : Nitrogen balances and protein requirements of
preterm infants. In : H.K.A. Visser, ed., Nutrition and
Metabolism of the fetus and infant, Martinus Nijhoff, The
Hague, Boston, London, 1979, p 195-212.

SUMMARY, GENERAL DISCUSSION AND SOME MATTERS OF THINKING

Alexandre Minkowski

I would like to try to pick up some of the major topics presented by the contributors and to raise questions, and problems.

Hepatic glycogen : enzymatic control.
Dr HERS reminds us the existence of 2 mechanisms of glycogenolysis, one by phosphorylation and the other one, especially interesting, by hydrolysis. The latter, lysosomial in essence, is associated with autophagic cellular digestion. One has to distinguish them between the "early labelling" with a delayed plateau whereas the "late labelling" displays at the same time a descending slope.

Hepatic glycogen : hormonal control.
Dr. MOSES - Reference is made to the technics of liver explant and to fetals hepatocytes cultures (PLAS) used also in our Laboratory by Drs Roux and Fulchignoni.

The action of cortisol, which is well known in the rat fetus is supposed to be the same in the human fetus, but not actually demonstrated. This remains an unsolved question.

Christiane PLAS demonstrated a time depend insulin effect in the fetal rat hepatocytes which develop a resistance to glucose incorporation into glycogen and a decrease of glycogen content. This is still unexplained.

The glucagon effect in the fetal rat is striking.

In comparing the responses to glucagon in hepatocytes with in vivo experiments, it is obvious that, in hepatocytes, a glucagon dependent pathway is present early in development before any definite storage of glycogen occurs.

More extensive studies should be undertaken in the developmental pattern of some non cyclic AMP pathways of phosphorylase activation which are known in the adult animal.

Similarities between glycogen metabolism in the rat and the human are likely, but with differences in time and magnitude.

Dr DE MEYER reviews the question of extrahepatic glycogen stores. He points out that 50 % of the stores are accumulated in the muscle, that an enormous accumulation takes place in the rat from day 15 to day 20.

Barbiturates and starvation lower the liver stores but not the extrahepatic stores.

The same phenomenon is observed under the action of adrenal depletion by reserpin.

Shelley notices that extrahepatic stores are even far more important in the human fetus.

There is no explanation in the presence of glycogen in the GI tract and the non endocrine pancreas in the human during the first half of gestation. This glycogen disappears later.

Minkowski, about the talks by Dr De Meyer and Moses, raises the question of the meaning of studying the fetus alone without the placenta.

Dr JONES - According to the speaker, blood glucose in the fetus is predominantly controlled by the transfer of glucose across the placenta and secondarily by insulin and less by catecholamines and glucagon.

It is of interest to study the problem using other species. In the guinea pig and the rabbit, the normal glycogen turnover is different from that observed in the rat.

In IUGR fetuses, glucagon is much lower than normal.
IUGR fetuses (rabbits and guinea pigs) have not <u>necessarily</u>
low glycogen hepatic concentration.   There are deffective
glycogen pathways in IUGR rabbits and guinea pigs fetuses
whereas not in the rat (Roux).

In experiments on perfused liver, metabolite concentration is very different in a 29 day old rabbit fetus and
in a 6 day postnatal animal.

For Minkowski, this again raises the problems of the
mechanism of postnatal adaptation which are not very clear
at the moment.

<u>Dr SNELL</u> in studying the turnover of glucose in newborns,
uses C^{14} or H^3 glucose.

Minkowski raises the question of using C^{13} glucose,
a stable non ionised glucose, giving information on the
molecule and the atome.

Snell shows that, after birth, there is a fall both in
plasma glucose and lactate.

Minkowski raises the question of the reconversion of
lactate into glucose.   How does it take place in pathological states such as hypoxia.

<u>Dr. ROUX</u> studies the carbohydrates reserves of the fetuses.

Glycogen is present in nearly all foetal tissues including : brain, kidney, gut; the maximum foetal level being
generally higher than the adult one.

Three general patterns have been described depending
on the organ.   Some organs display a maximum level at mid
gestation as in lung or around the three fourth as in placenta.   Skeletal muscle, liver accumulate glycogen till birth when maximal values are reached.   The cardiac glycogen
follows an inverse pattern, presenting its maximum early
in gestation and decreases continuously till birth.
Since fair amounts of glycogen are present in the placenta
and lung at a time when there is very little in the liver,
it has been suggested that these act as temporary glycogen

stores until the liver is able to take over.

It is, at this point, emphasized that little is known about the factors responsible for the first appearance of glycogen in the other fetal tissues and for the subsequent changes in concentration. A direct correlation between the level of enzymes, quantified in vitro and the level of glycogen is hazardous. Some of the enzymes have to be activated and the regulation of their activity, "in vivo" is under the strict dependence of endocrine and metabolic factors.

Betamethasone 17,21 dipropionate inactive to induce glycogen deposition in the fetal liver in vivo and in vitro but suppresses the hormonal action of cortisol remains UNEXPLAINED.

Corticosteroids of maternal origin are extensively converted to their biologic inactive II-dehydroderivatives in the placenta and fetal tissues.

Conversion of dexa and betamethasone is negligible.

Rapid exchange of cortisol and prednisolone takes place across the placenta in both directions.

The role of hormones is entirely dependent on the timing during gestation.

In IUGR rats (10 years ago) Roux and others reported that enzymes for glycogen synthesis in the fetus were not impaired.

Only few information on glycogen reserves in the human fetus are actually available.

Dr FERRE - Factors affecting glucose metabolism in the newborn.

At birth major nutritional changes take place whereas the foetus oxidizes primarily glucose and amino acid, the newborn rat gets a lipid rich diet : milk. (70 % of the total caloric intake in the rat milk derives from lipid sources and only 10 % from carbohydrates).

The amount of glucose provided by milk daily (15 mg) represents only 10 % of this rate of renewal. The newborn is then compelled to produce de novo the majority of its gluconeogenesis.

In injecting to the suckling I day rat 3-mercaptopico-
linate, an inhibitor of PEP carboxykinase, one produces a
deep lowering of gluconeogenesis, followed by a marked
hypoglycemia.

Gluconeogenesis measured in isolated hepatocytes of
suckling newborn rats shows that it is comparable to that
of the fasting adult rat and 5 times higher than in the fed
adult rat.

Hepatic gluconeogenesis, high in the suckling animal
is very low on the fasting animal.

Key-enzymes of gluconeogenesis, pyruvate carboxylase,
PEP carboxykinase, fructose diphosphatase and G6Pase are
comparable in both groups as well as hormonal environment.

On another hand in the fasting newborn rat, very low
concentration of glucoproducing substances are found as
well as substrates derived from lipids, as compared to
suckling animals.

Injection of glucoproducing substrates in the fasting
newborn produces concentrations comparable to the ones
observed in suckling animals.

Fasting rats given triglycerides have concentrations
of fatty acids, ketonebodies and glycerol comparable to
those of suckling animals.

In the newborn rat, hepatic oxidation of fatty acids
is necessary for an active gluconeogenesis.

Comments by A. MINKOWSKI.

It is also necessary to study animals with high deposition of fat in the fetus (bat, guinea pig). Metabolic
responses are likely to be different and are more comparable with the human (high deposit of fetal fats).

Dr KOLLEE in IUGR animals shows that it is interesting to compare SHAM operated animals to animals of the non ligated
horn.

SHAM operated have higher glucose than animals of the

ligated horn (IUGR) and less than controls (non ligated horn).

(Alanine, glycine and threonine are less elevated in SHAM than in IUGR).

In the general discussion of the previous papers, Girard pointed out that the rabbit undergoes 48 hours starvation without hypoglycemia.

Dr HULL insists on the transfer of lipids through the placenta, he drew attention on the utilisation of I.V. lipid preparations. He also raised the question of the depletion in a specific adipose tissue (brown fat more than white) and stresses the importance of fat stores in the liver.

Shelley emphasizes the need of measuring blood flow.

Dr VERELLEN and Dr SAUER showed figures on caloric needs. Senterre disagrees on the figures of N retention in breast fed and formula fed babies. The real cost of growth is not actually represented.

Milner comments on the composition of expressed, pumped milk.

Minkowski reminds of the work by Baum (Oxford) in which it is shown that human milk composition varies during feeding : there is an enormous increase of fat content in a few minutes.

Dr HULL shows that there are enormous variations in O_2 consumption with postnatal age, after feeding, sleep state, motion.

Individual variations are huge. Local variations are reflected by 32° in the foot, 37° in ear and vectors. Hull measures enormous transepidermal water loss by evaporimeter (20 - 50 % of total heat loss). This can be partially prevented by paraffin ointment.

Dr SAUER studies the cost of growth. Different results are obtained by indirect and direct calorimetry. These findings are discussed by Senterre.

Small for date infants retain less fat than preterm infants.

Dr BOSSI shows a very fine model to study energetic substrates in brain cell cultures (with only very few neurones).

Dr SODOYEZ gives a perfect highlight on insulin secretion. Fetal blood glucose and fetal plasma insulin do not correlate. High amounts of amino acid levels in the fetus might play a role in high insulin levels. Study in pups with and without placenta (insulin highest when placenta remains connected, and glucose higher when placenta is withdrawn) emphazises the role of a placenta-insulin axis.

About the paper of Dr GIRARD on glucagon secretion.

The question is raised : is it really glucagon which is measured by radio immuno-assay.

According to GIRARD, under hypoxic conditions, lactate raises, insulin does not change despite change in glucose.

Metabolic response to maternal fasting is characterized by lowering of plasma insulin and increase of plasma glucagon in the fetus.

The same is observed in prolonged pregnancy, and after injection of phloridzin.

SHELLEY notices that in a sheep chronic preparation : fasting hypoglycemia raises insulin level in the fetal lamb but does not raise glucagon level. Glucagon changes in hypoxia with high modification of the pH.

MILNER administers 1 unit protaminzine insulin to 27 day old rabbit fetuses and measures somatomedin like activity by S 35 uptake. Enormous variations intra litter and from one litter to the other are observed but nevertheless it is possible to demonstrate that insulin has an anabolic effect on skeleton and cartilage.

SALLE shows an elevation of somatomedin in the premature, but not in the S.G.A. In cord blood, no a/v differen-

ce in premature and S.G.A. but significant a/v differences
in the full term are observed. Correlation between growth
hormone and somatomedin activity are discussed.

VAN ASSCHE makes an excellent review of diabetes and
pregnancy.
Among the questions, MINKOWSKI notices :
- very little knowledge on IUGR from diabetic mothers
- very long term effect of maternal diabetes :
  1) are changes in the newborn reversible after birth?
  2) Is fibrosis of the islet of Langerhans responsible
  for the incidence of juvenile diabetes (20 times higher)?
VAN ASSCHE shows in 1979 that mild experimental diabetes
causes persistent changes in endocrine pancreas of their
offspring (second generation). This may be interpreted as
a possible cause of gestational diabetes, which is even
manifest in the fetuses of the third generation.
MINKOWSKI quotes the recent issue of CIBA SYMPOSIUM 1979,
"Pregnancy, Metabolism, Diabetes in the mother and the
fetus", Excerpta Medica, Pub.

The discussion on oral feeding.
It raises the point of how much protein intake is preferable for the premature infant.
     RÄIHIÄ's work shows that mother's milk provides a low
amount (2gr/kg/day) of protein which prevents high amounts of tyrosine
and phenylamine (as observed with cow's milk), potentially
toxic for the brain.
     In SENTERRE's view the premature infant fed with human
milk is in hypoaminoacidemia and should retain more nitrogen.
The protein intake then should progressively reach 3.5g/kg/
day.

In the discussion on feeding some points are raised.
1) About the use of fructose, SALLE notices that when fructose is given, insulin is not necessary and hyperglycemia
does not occur.

DE MEYER points out that fructose should be given with some
reservation as high level of fructose blocks aldolase.
2) About the very low level of plasma amino-acids when the
premature is given only mother milk, SENTERRE notices that
it is lower than the amino-acid level in the cord blood.
He points out that one should analyse each amino acid every
day because of large daily changes.
The levels change with the degree of maturity.
3) As the protein intake with mother milk is too low, one
should enrich this milk with protein up to 3.5 g/kg/day in
the premature weighing less than 1 kg. But it is always
necessary to keep some mother milk in the diet.
When DE MEYER asks if it is necessary to give casein,
SENTERRE says that it is possible to do with hydrolysates.
4) Is it necessary to add insulin?
SENTERRE replies : 1 microunit/kg/hour, but SALLE states
that this is dangerous.

At the end, MINKOWSKI raises his own questions and suggests
some points that merit more investigation.
A) It is important to study the relation between placenta
and the fetus.
B) The problem of conversion of lactate to glucose in patho-
logic states has to be investigated.
C) Study in animals with large fetal deposit of fat must
be performed.
D) The effect of the quality and the quantity of food upon
brain development is particularly important. In this
regard, SENTERRE's views are different from RÄIHIÄ's.
E) It seems impossible to separate the role of carbohydrates
from other nutrients.
F) Multifactorial studies are necessary.
G) The role of fructose has to be defined more closely.